Just the Facts: Essentials of Environmental Science

First Edition

© 2021 National Agricultural Institute, LLC.

ALL RIGHTS RESERVED. No part of this work covered by the copyright herein may reproduced except as permitted under Section 107 or 108 of the United States Copyright Act, without the prior written permission of the publisher.

First Edition, August 2021

National Agricultural Institute, LLC
151 W. 100 S.
Rupert, ID 83350
USA
(208) 670-3704

ISBN: 978-1-4951-7129-1

To learn more about the National Agricultural Institute, LLC visit:

www.national-ag-institute.org
Or visit us on Facebook: www. facebook.com/Agri101

Send comments or questions to roparker@national-ag-institute.org

Notice to the Reader

Publisher does not warrant or guarantee any of the products described herein or perform any independent analysis in connection with any of the product information contained herein. Publisher does not assume, and expressly disclaims, any obligation to obtain and include information other than that provided to it by the manufacturer. The reader is expressly warned to consider and adopt all safety precautions that might be indicated by the activities described herein and to avoid all potential hazards. By following the instructions contained herein, the reader willingly assumes all risks in connection with such instructions. The publisher makes no representations or warranties of any kind, including but not limited to, the warranties of fitness for particular purpose or merchantability, nor are any such representations implied with respect to the material set forth herein, and the publisher takes no responsibility with respect to such material. The publisher shall not be reliable for any special, consequential, or exemplary damages resulting, in whole or part, from the readers' use of, or reliance upon, this material.

Table of Contents

1. Introduction 1
2. Global Environmental Issues, Causes, Responses, and Future 5
3. An Approach to Dealing with Environmental Issues 7
4. Resources 10
5. Chemical Concepts 15
6. Life – Elements, Energy, and Matter 20
7. Photosynthesis 25
8. Cellular Respiration 28
9. An Ecosystem and How it Works 31
10. The Principles of Ecosystem Functions 44
11. Ecosystem Balance and Imbalance 49
12. Ecological Succession 55
13. Ecosystem – Adapting to Changes 58
14. Wildlife & Wildlife Management 64
15. Pollution 69
16. Water, Water Cycle & Water Purification/Treatment 73
17. Eutrophication 82
18. Sewage Pollution 85
19. Removing Pollutants from Sewage 90
20. Major Air Pollutants and Their Impacts 95
21. Soil and Soil Ecosystem 98
22. Pests and Pest Control 104
23. Global Human Population 107
24. Pollution from Hazardous Chemicals 113
25. Major Environmental Laws on Hazardous Wastes 117
26. Sustainability Concepts 120
27. Basics of Composting 139
28. Landfill Gas Energy Basics 149
29. Recycling 165
30. Entomophagy 191

Glossary 199

Preface

This textbook, *Essentials of Environmental Science*, is one in a series of *Just the Facts (JTF)* textbooks created by the National Agricultural Institute. This is a bold, new approach to textbooks. These textbooks present the essential knowledge in outline format. This essential knowledge is supported by a main concept, learning objectives and key terms at the beginning of each section. Content of the books is further enhanced for student learning by connecting with complementary PowerPoint presentations and websites through QR codes (scanned by smart phones or tablets) or URLs. Each textbook is available in print and electronic formats.

The time is now for a new mindset about textbooks. Textbooks for the future need to take advantage of both print and digital technology, while keeping costs down.

Just the Facts series of textbooks provides a synergistic textbook model - print and digital working together to be better than either one alone. Moreover, in a time of increasing costs for textbooks, print copies of Just the Facts textbooks are $50 or less per book.

The first of these new textbooks also includes:

- *Just the Facts: Introduction to Agriculture*
- *Just the Facts: Introduction to Agribusiness*
- *Just the Facts: Introduction to Animal Science*
- *Just the Facts: Introduction to Biology*
- *Just the Facts: Introduction to Food Science & Food Systems*
- *Just the Facts: Introduction to Plant Science*
- *Just the Facts: Introduction to Soil Science*

Just the Facts textbooks are a project of the National Agricultural Institute, LLC created, written and assembled by:

Rick Parker, PhD, President
Marilyn Parker, Vice President
Karen Kenny, Assistant Editor
Dylan Stott, Assistant

For more information check out our website: www.national-ag-institute.org
Or email: info@national-ag-institute.org

1 Introduction

Major Concept

Environmental science is the branch of science concerned with environmental problems and issues. It is the study of how species interact with one another and with the nonliving environment.

Objectives

- Compare an environmental scientist to an environmentalist
- Describe an environment
- Identify the difference between conservationist, preservationist and cornucopians
- Define biosphere and identify the three spheres contained within it
- Identify two abiotic and two biotic resources
- Describe sustainability
- List the ecosystems of the Earth

Key Terms

- Abiotic
- Atmosphere
- Biodiversity
- Biosphere
- Biotic
- Conservation biologists
- Conservationist
- Cornucopian
- Development
- Ecosphere
- Ecosystems
- Environment
- Environmental Science
- Environmentalist
- Hydrosphere
- Lithosphere
- Preservationist
- Resource
- Science
- Sustainability
- Sustainable development
- Sustainable yield
- Technology

Chapter Resource

Complementary *full color* illustrations, photos, charts, and graphs are available for Chapter 1 by following this URL: https://tinyurl.com/7vxzdaks This digital resource will enhance your understanding of the chapter concepts.

Environmental Science

- **Science** is any systematic field of study involving experiment, observation, and deduction, to produce reliable explanation of phenomena, with reference to the material and physical world.

  o May be divided into separate areas of study, such as:

    ✓ Astrology
    ✓ Biochemistry
    ✓ Biology
    ✓ Chemistry
    ✓ Ecology
    ✓ Environmental science
    ✓ Geography
    ✓ Geology
    ✓ Life science
    ✓ Mathematics
    ✓ Physics
    ✓ Statistics
    ✓ Technology

- **Technology** is applied science, which results in creation of new products and processes intended to improve our efficiency, chances for survival, level of comfort and quality of life.

    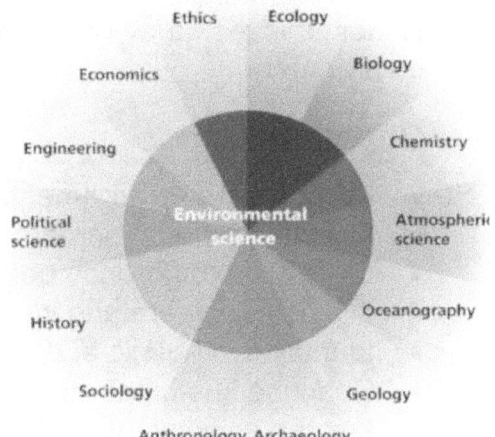

    o Science and technology are highly organized fields of study.

    o Technological innovations can cause and solve many of the environmental problems/issues that we face today.

- An **environment** can be explained as the physical surroundings and factors.

    o Living things (organisms)
    o Nonliving things (matter and energy)

        ✓ These factors may affect natural systems during their lifetime.

- **Environmental science** is the branch of science concerned with environmental problems and issues. It is the study of how species interact with one another and with the nonliving environment.

    o Environmental science is simply a study of connections and interactions in nature.

    o Integrates knowledge from wide range of disciplines including:
        ✓ Biology (specifically ecology)
        ✓ Chemistry
        ✓ Demography
        ✓ Economics
        ✓ Ethics
        ✓ Geography
        ✓ Geology
        ✓ Physics
        ✓ Politics
        ✓ Psychology
        ✓ Resource conservation and management
        ✓ Resource technology and engineering
        ✓ Sociology

Conservation and Preservation

- Environmental scientists use the scientific method, data collection analysis and information to understand how the earth works, learn how humans interact with the earth, and develop solutions to environmental problems.

- Any person who is concerned with the protection of the environment or who believes that sustainability of civilization depends on conserving natural aspects of the **biosphere** free from pollution and maintaining **biodiversity** is referred to as an **environmentalist**.

- Those who create multidisciplinary science to investigate human impacts on the diversity of life (biodiversity) on earth and develop practical strategies or plans for preserving biodiversity are **conservation biologists**.

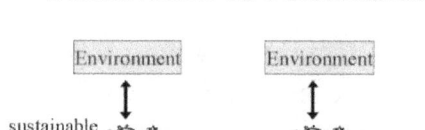
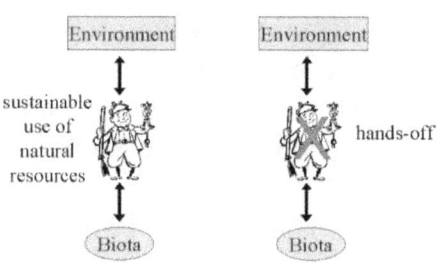

Conservation vs. Preservation

- **Conservationists** are concerned with using natural areas and wildlife for the benefit of present and future generations of human races and other forms of life.

- Those concerned with the preservation of natural areas are called **preservationists**.

    o Primary goal is to ensure that undisturbed natural areas are free from harmful human activities.

- **Cornucopians** are those who assume or believe that all parts of the environment are natural resources to be exploited for the advantage of humans.

Biodiversity and Sustainability

- **Biosphere** (also known as **ecosphere**) is the portion of the earth where life is found (overall **ecosystems** of the Earth).

    o Consists of:

    ✓ **Atmosphere** (the troposphere)
    ✓ **Hydrosphere** (surface water and ground water)
    ✓ **Lithosphere** (soil, surface rocks and sediments on land; bottom of oceans and bodies of water, where life is found)

    o Sum total of ecosystems.

    ✓ All interconnected and independent through global processes.

- **Biodiversity** refers to the diversity (variety) of living things or species diversity (biological resources) found in the natural world.

    o Usually refers to the different species, but also includes a variety of ecosystems and the genetic diversity within a given species.

- A **resource** refers to anything obtained from **biotic** (living) and **abiotic** (nonliving) environment to meet human needs and desires.

    o Resources can be sustainable or unsustainable.

- **Sustainability** is the ability of a system to survive for some specified time without depleting the energy or material resources on which it depends.

- **Sustainable development** is a form of economic growth and ecological activities that do not deplete or degrade natural resources on which present and future economic growth and life depend.

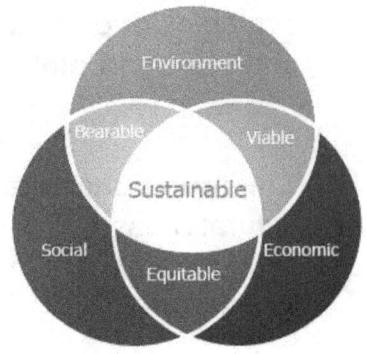

  - The development that provides people and other living organisms with a better life without sacrificing or depleting resources or causing environmental impact that undercut future generations.

  - **Development** (progress) refers to a stage of growth or advancement in using resources. This can be sustainable or unsustainable.

- The highest rate at which a potentially renewed resource can be used without reducing its available supply in a particular area or region is known as the **sustainable yield**.

  - The taking of a biological resource (e.g., forest) that does not exceed the capacity of the resource to reproduce and replace itself.

Additional Resources

Internet sites represent a vast resource of information. Using one of the search engines on the Internet such as Google or DuckDuckGo, find more information by searching for these words or phrases: Always use caution when searching for information on the Internet. Follow guidelines to ensure the accuracy and reliability of the information you find.

Search words: Environmental science, biosphere, environmentalist, biodiversity, conservation biologists and sustainable development.

Self-Assessment

1. Choose an environmental science discipline and describe how it integrates with the environment.

2. How would you describe your own environmental beliefs? And why?

3. Describe the biodiversity in your local area.

4. Identify a sustainable development project in your area and write a brief explanation.

2 Global Environmental Issues, Causes, Responses, and Future

Major Concept

Environmental issues need to be categorized and identified before seeking solutions.

Objectives

- Name four categories and four causes of environmental issues.
- Identify how humans may respond to environmental issues.

Key Terms

- Ozone
- Poaching

Chapter Resource

Complementary *full color* illustrations, photos, charts, and graphs are available for Chapter 2 by following this URL: https://tinyurl.com/7vxzdaks This digital resource will enhance your understanding of the chapter concepts.

Environmental Issues

- Environmental issues may be categorized as:
    - Air pollution
    - Water pollution
    - Soil erosion and pollution
    - Coastal and marine pollution and degradation
    - **Ozone** depletion
    - Climate change
    - Land degradation
    - Deforestation and habitat loss
    - Loss of biodiversity
    - Environmental hazards
    - Toxic chemicals and hazardous/solid wastes

- Causes of environmental issues include:
    - Agriculture
    - Fisheries
    - Industries
    - Energy
    - Transport
    - **Poaching**
    - Tourism
    - Human settlement
    - Population
    - Health
    - War
    - Peace
    - Security

- Human responses to environmental issues may include:

    - Understanding the environment
    - Perceptions and attitudes towards the environment
    - International/national/local/individual responsibilities

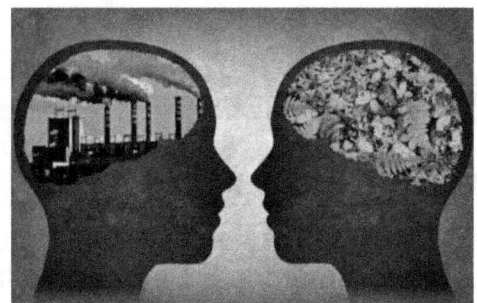

- Looking into the future

    - Challenges and opportunities

        ✓ Understanding environmental and developmental interrelationships
        ✓ Practicing interdependence

Additional Resources

Internet sites represent a vast resource of information. Using one of the search engines on the Internet such as Google or DuckDuckGo, find more information by searching for words or phrases: Always use caution when searching for information on the Internet. Follow guidelines to ensure the accuracy and reliability of the information you find.

Search words: Global environmental issues
Websites: www.NASA.gov, www.unep.org

Self-Assessment

1. What do you feel is the most important global environmental issue and why?

2. What do you think contributes most to the causes of environmental issues?

3 An Approach to Dealing with Environmental Issues

Major Concept

To determine an environmental or an ecological problem/issue, to pose questions to and then try to find possible answers.

Objectives

- Define an environmental or ecological issue

Chapter Resource

Complementary *full color* illustrations, photos, charts, and graphs are available for Chapter 3 by following this URL: https://tinyurl.com/7vxzdaks This digital resource will enhance your understanding of the chapter concepts.

Determining Environmental Problems/Issues

- In determining whether there is an environmental or an ecological problem/issue, you need to pose questions to yourself first and then try to find possible answers.

    o You may interview knowledgeable people or those affected by the problem/issue.
    o You will need to define and redefine the problem/issue with the help of those concerned.
    o Then plan for actions to be implemented by those affected and if possible, with the help from outside.

- What is a problem/issue?

    o The word problem or issue means different things to different people.

        ✓ It can be a frustrating situation beyond our control.
        ✓ It can also be intricate, unsettled questions that seemingly defy understanding or questions that call for action.

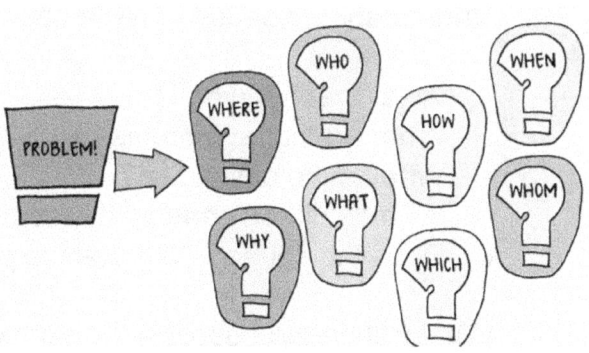

- Defining the problem/issue

    o What is the problem/issue?
    o What caused the problem/issue?
    o Is this my problem/issue?
    o Do I have the authority to address the problem/issue?
    o In what context will I address the problem/issue?
    o What is the first step?

Questions that Call for Action

1. Making a decision

    - What decision do you need to make?
    - What do you want to achieve?
    - What are the short-term objectives?
    - What are the long-term goals?
    - On what values will the decision be based?
    - What criteria flow out of these values?
    - What alternative solutions exist?
    - What are the consequences of each alternative relative to each criterion?
    - Which alternative will most effectively achieve your objectives consistent with your long-term goals?

2. Planning action

    - What needs to be done?

        - ✓ By whom?
        - ✓ When?

    - After each step is acted on, what was done?

        - ✓ By whom?
        - ✓ When?

    - Did the action taken achieve the intended objectives?

    - If nothing was done to address the problem, what were the reasons?

        - ✓ Can it be done now?
        - ✓ Why and how?

3. Understanding cause and effect

    - What are the symptoms?
    - What are not symptoms?
    - What is the difference?
    - What could be causing the symptoms?
    - Does the cause adequately explain the symptoms?

4. What will be the effect on the economy and on social issues?

    - Some fixes of issues destroy the economy in the area. Can this be justified?
    - Some fixes to environmental issues come with a huge social cost.

5. How much will it cost?

    - Can the cost be justified? Anything can be done if you have the money to spend, but is the dollar cost worth it?
    - Will any of the money spent be recovered in some real savings?

6. Is the issue real or perceived?

    - In many cases, perception becomes reality. Would it be better to change the perception of some issue or is it possible?

Assignment

Choose an environmental issue and find answers for each of the six questions above and present your answers in a 3-minute Pecha Kucha.

Additional Resources

Internet sites represent a vast resource of information. Using one of the search engines on the Internet such as Google or DuckDuckGo, find more information by searching for words or phrases: Always use caution when searching for information on the Internet. Follow guidelines to ensure the accuracy and reliability of the information you find.

Websites: www.NASA.gov or www.ncei.noaa.gov

4 Resources

Major Concept

Materials obtained from the living organisms and nonliving environment to meet human needs and desires are resources.

Objectives

- Provide examples of ecological and economic resources
- Explain renewable, potentially renewable, and nonrenewable resources
- Describe the concept of sustainability
- Know what sustainable living means

Key Terms

- Biological diversity (biodiversity)
- Ecological diversity
- Ecological resources
- Economic resources
- Environmental degradation
- Exhaustible
- Genetic diversity
- Nonrenewable
- Potentially renewable
- Renewable resources
- Species diversity
- Sustainable living
- Sustainable yield

Chapter Resource

Complementary *full color* illustrations, photos, charts, and graphs are available for Chapter 4 by following this URL: https://tinyurl.com/7vxzdaks This digital resource will enhance your understanding of the chapter concepts.

Resources

- **Resources** are materials obtained from the living organisms and nonliving environment to meet human needs and desires.

- They can be ecological or economic resources.

    o **Ecological resources** are anything required by living organisms for the purpose of normal maintenance, growth, and reproduction processes. Examples include:

    - ✓ Habitat
    - ✓ Food
    - ✓ Water
    - ✓ Shelter

    o Economic resources are anything obtained from the environment to meet human needs and desires (essential or non-essential). Examples include:

    - ✓ Food
    - ✓ Water
    - ✓ Shelter
    - ✓ Manufactured goods
    - ✓ Transportation
    - ✓ Communication
    - ✓ Recreation

- On human short-term perception scale, we categorize the material resources we get from the environment as renewable, potentially renewable, and nonrenewable resources.

- Some resources are directly available for use, such as solar energy, fresh air, wind, fresh surface water, fertile soil, and wild edible plants.

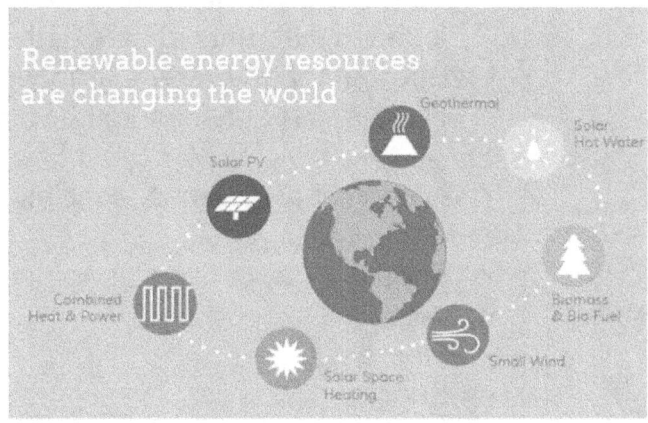

- Others such as petroleum (oil), iron, groundwater, and modern crops are not directly available.

    o They become useful to us only with some effort and technological inventiveness.

        ✓ For example, petroleum was a hidden fluid until we learned how to find, extract, and convert (refine) it into gasoline, heating oil and other products that could be sold at affordable prices.

- Human struggle for resources is common and sometimes lead to conflicts including wars, civil war, ethnic conflicts, or even individual fights.

Renewable, Potentially Renewable, and Nonrenewable Resources

- Resources that are continually replenished on the human time scale are considered **renewable** or **everlasting resources**.

    o Solar energy is a good example of a renewable resource. It is expected to last at least 6 billion years as the sun completes its life cycle.

- **Potentially renewable** resources can be replenished rapidly (hours to several decades) through natural processes.

    o Examples are:

        ✓ Air
        ✓ Fertile soil
        ✓ Forests
        ✓ Fresh lake and stream water
        ✓ Grasses
        ✓ Groundwater
        ✓ Wild animals

    o An important potentially renewable resource is **biological diversity** or **biodiversity**: the different life forms (species) that can best survive the variety of conditions currently found on the earth. These include:

- ✓ **Genetic diversity** - variety of the genetic makeup among individuals within a species
- ✓ **Species diversity** - variety among the species or distinct types of living organisms found in different habitats of the planet.
- ✓ **Ecological diversity** - variety of forests, deserts, grasslands, streams, lakes, oceans, wetlands, and other biological communities.

o The rich variety of genes, species, and biological communities give us:

- ✓ Food
- ✓ Wood
- ✓ Fibers
- ✓ Energy
- ✓ Raw materials
- ✓ Industrial chemicals
- ✓ Medicines

o These all pour hundreds of billions of dollars into the world economy each year.

o Potentially renewable resources can be depleted.

o The highest rate at which a potentially renewable resource can be used indefinitely without reducing its available supply is called **sustainable yield**.

o If we exceed a resource's natural replacement rate, the available supply begins to shrink. This is known as **environmental degradation**. Examples include:

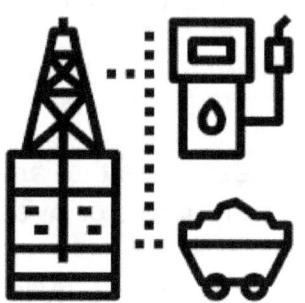

- ✓ Urbanization of productive land
- ✓ Waterlogging and salt buildup in soil
- ✓ Excessive erosion on topsoil
- ✓ Deforestation
- ✓ Depletion of ground water
- ✓ Overgrazing of grasslands by livestock
- ✓ Loss of biodiversity caused by elimination of habitats and species
- ✓ Pollution caused by industries, automobiles, fertilizers, pesticides, fires, urban and industrial wastes, and waste treatment plants

- **Nonrenewable resources** are resources that are present in a fixed amount (in stock) in various places in the earth's crust and have the potential for renewal only by geological, physical, and chemical processes taking place over hundreds of millions to billions of years.

    o Examples include:

    - ✓ Energy resources (coal, oil, natural gas, uranium) which cannot be recycled

- ✓ Metallic mineral resources (copper, aluminum, iron) which can be recycled
- ✓ Nonmetallic mineral resources (salt, clay, sand, and phosphates) which are usually difficult or expensive to recycle

  o We categorize these resources as **exhaustible** because we extract and use them at a much faster rate than they are being formed.

Concept of Sustainability

- The ability of a system to survive and function over some specific time is referred to as **sustainability**.

  o The system satisfies the needs of its inhabitants without depleting natural capital and thereby jeopardizing the prospects of current and future generations of humans and other species.

- Sustainable living is the process of taking no more potentially renewable resources from natural world than can be replenished by natural processes.

  o For example, our existence, lifestyles, and economies depend completely on the sun and the earth. To economists, capital is wealth used to sustain a business and to generate more wealth. We can think of energy from the sun as solar capital and the planet's air, water, soil, wildlife, minerals, and natural purification, recycling, and pest control processes as natural resources or natural capital.

- Living sustainably means living on income by not utilizing the capital that supplies the income.

  o Imagine that you inherit $1 million. Invest this capital at 10% interest per year and you will have a sustainable annual income of $100,000 (you can spend up to $100,000 a year without touching the capital that sustains your lifestyle).

Additional Resources

Internet sites represent a vast resource of information. Using one of the search engines on the Internet such as Google or DuckDuckGo, find more information by searching for words, phrases, or websites: Always use caution when searching for information on the Internet. Follow guidelines to ensure the accuracy and reliability of the information you find.

Websites:
www.NASA.gov
www.ncei.noaa.gov

101 Terrific Sites on Renewable Energy
https://www.environmentalsciencedegree.com/terrific-renewable-energy/

Self-Assessment

1. List four resources that are directly available for use by humans.

2. Identify an example of environment degradation and the potential impact on the environment.

3. Compare and contrast two potentially renewable and nonrenewable resources.

4. Explain sustainable living.

5 Chemical Concepts

Major Concept

Basic chemistry is required in understanding how an ecosystem works.

Objectives

- Identify the structure of an atom
- Understand how the number of protons and electrons determine the chemical property of an element
- Explain the importance of covalent bonding
- Describe the process of ionic bonding

Key Terms

- Atom
- Compound
- Covalent bond
- Electrons
- Element
- Ion
- Ionic bond
- Matter
- Mixture
- Molecule
- Neutrons
- Nucleus
- Protons

Chapter Resource

Complementary *full color* illustrations, photos, charts, and graphs are available for Chapter 5 by following this URL: https://tinyurl.com/7vxzdaks This digital resource will enhance your understanding of the chapter concepts.

Chemical Basis of Life

- Basic chemistry is required in understanding how an ecosystem works. An ecosystem is made up of the living and nonliving environments.

  o These environments are made up of matter. Matter may be made through chemical reactions. Energy makes it possible for chemical reactions to take place.

  o **Matter** can be defined as anything that occupies space and has mass (gas, liquid, or solid).

States of Matter

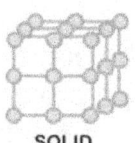  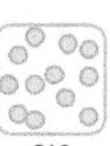
SOLID LIQUID GAS

  o Matter is made up of tiny units called atoms. An **atom** is the basic unit of matter – the smallest particle of an element that maintains its chemical identity through all chemical and physical changes.

- Atoms bond together to form different compounds. Atoms of one kind can join to form elements.

  o An **element** is a substance that is made up of one and only one distinct kind of atom.

- ✓ Elements cannot be created and destroyed in a chemical and biological reaction. They cannot be changed from one to another.

- o Atoms can bond together to form molecules. A **molecule** is a specific union of two or more atoms bonded together – the smallest unit of a compound that still has characteristics of that compound.

    - ✓ A **compound** is any substance that is made up of two or more different kinds of atoms bonded together.

- The term **mixture** in chemistry means there is no chemical bonding between the molecules of the elements involved.

    - o For example, air is a mixture of oxygen, nitrogen, carbon dioxide, and rare gases.

Atoms, Bonds, and Chemical Reactions

- In chemical reactions, atoms are neither created, nor destroyed, nor changed to another kind of atom.

- What occurs in chemical reactions, whether mild or explosive is simply a rearrangement of the ways in which the atoms involved are bonded together.

    - o An oxygen atom, for example, may be combined and recombined with different atoms to form any number of different compounds, but the given oxygen atom always has been, and always will be, an oxygen atom. The same can be said for all the other kinds of atoms.

- To understand how atoms may bond and rearrange to form different compounds, it is necessary to learn the structure of atoms.

Structure of Atoms

- At atom consists of a central core called the **nucleus**. The nucleus of an atom contains one or more **protons** and, except for hydrogen, one or more **neutrons** as well.

- Surrounding the nucleus are particles called **electrons**. Each proton has a positive (+) electric charge, and each electron has an equal but negative (-) electric charge.

    - o Thus, the charge of the proton may be balanced by an equal number of electrons making the whole atom neutral. Neutrons have no charge.

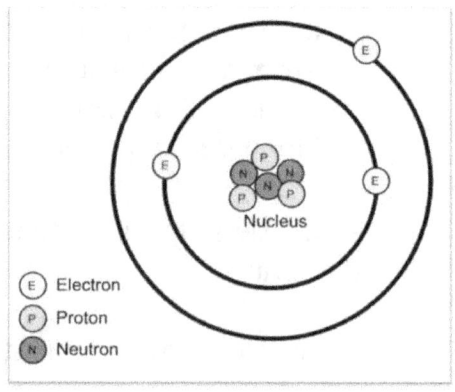

- Atoms of all elements have the same basic structure consisting of protons, electrons, and neutrons. The distinction among atoms of different elements is in the number of protons.

    - The atoms of each element have a characteristics number of protons, which is known as the atomic number of the element.

    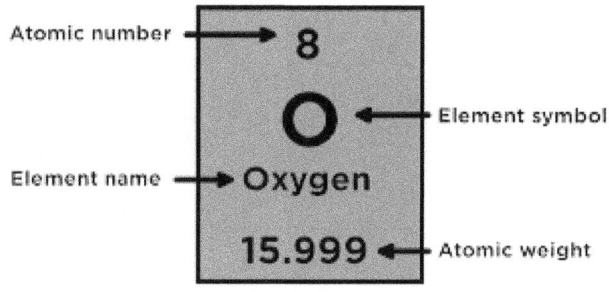

    - The number of electron characteristics of the atoms of each element also differs corresponding to the number of protons.

- The number of protons and electrons, i.e., the atomic number of the element, determines the chemical properties of the element. However, the number of neutrons may also vary.

    - Most carbon atoms have six neutrons in addition to the six protons. However, some carbon atoms have eight neutrons.

        - ✓ Atoms of the same element, which have different numbers of neutrons, are known as **isotopes** of the element.

Bonding of Atoms

- The chemical properties of an element are defined by the ways in which its atoms will react and form bonds with other atoms.

- By examining how atoms form bonds, we shall see how the number of electrons and protons determines these properties: (a) covalent bonding and (b) ionic bonding.

    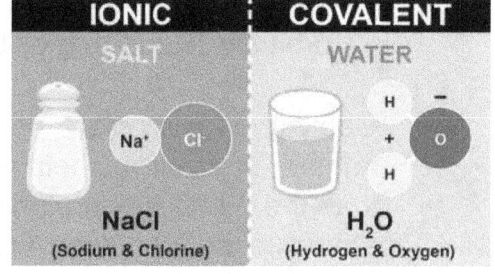

    - In both kinds of bonding, it is important to recognize that electrons are not randomly distributed around the atom's nucleus. Rather, there are, in effect, specific spaces in a series of layers, or orbitals or shells, around the nucleus.

    - If an orbit or shell is occupied by one or more electrons but not filled, the atom is unstable.

        - ✓ It will tend to react and form bonds with other atoms to achieve greater stability.
        - ✓ A stable state is achieved by having all the space in the orbit or shell filled with electrons.
        - ✓ It is also important to keep the charge neutral, i.e., the total number of electrons equals that of the protons.

- Adjacent atoms sharing one or more pairs of electrons may satisfy filling all the spaces and keeping the charge neutral. The sharing of a pair of electrons holds the atoms together in what is called a **covalent bond**.

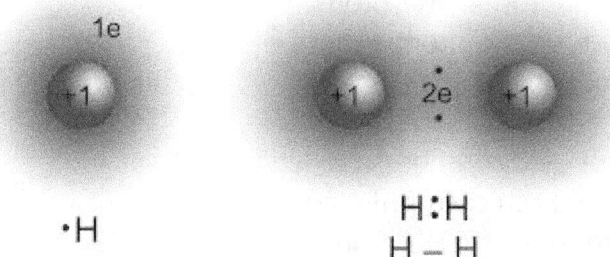

  o By satisfying the charge-orbital requirements, that leads to a discrete unit of two or more atoms bonded together. Such units of two or more covalently bonded atoms are called **molecules**.

- Another way in which atoms may achieve a stable electron configuration is to gain additional electrons to complete the filling of an orbit (shell), or lose electrons, which are over a completed orbit.

- In general, the maximum number of electrons that can be gained or lost by an atom is three. Therefore, an element's atomic number determines whether one or more electrons will be lost or gained.

  o If an atom's outer orbit is one to three electrons short of being filled, it will tend to gain additional electrons.

  o Conversely, if an atom has one to three electrons over its last complete orbit it will tend to give them away.

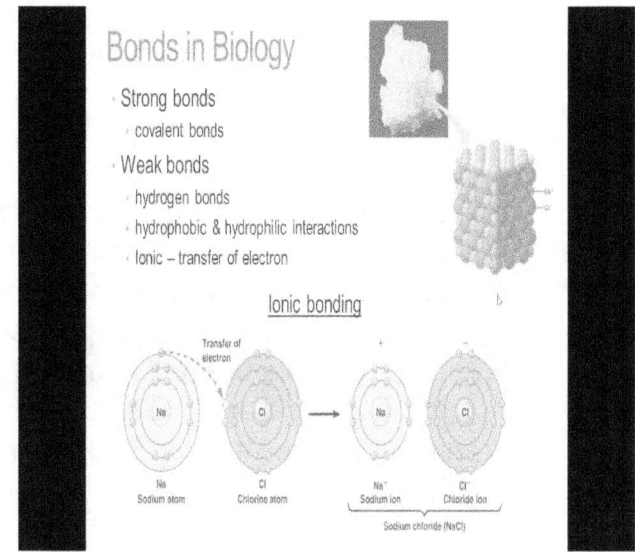

- Gaining or losing electrons results in the number of electrons being greater or less than the number of protons, and the atoms consequently having an electric charge.

  o The charge will be one negative for each electron gained or one positive for each electron lost.

  o An atom or group of atoms which has acquired an electric charge in this way is called an **ion**, positive or negative.

- Since unlike charges attract positive and negative ions, they tend to join and pack together in dense clusters in such a way as the neutralize the overall electric charge.
    - ✓ This joining together of ions through the attraction of opposite charges is called **ionic bonding**.

Additional Resources

Internet sites represent a vast resource of information. Using one of the search engines on the Internet such as Google or DuckDuckGo, find more information by searching for words, phrases, or websites: atoms, structure of atoms, covalent bonding, and ionic bonding. Always use caution when searching for information on the Internet. Follow guidelines to ensure the accuracy and reliability of the information you find.

Websites: Chapter 4: The Basics of Chemistry https://chem.libretexts.org/

Self-Assessment

1. List the three strong bonds and the one weak bond of atoms.

2. Draw and label an oxygen atom.

3. Describe what happens to atoms in a chemical reaction.

4. How is an element different from an atom?

6 Life – Elements, Energy, and Matter

Major Concept

Life can be defined by the combination a variety of characteristics involving elements, energy, and matter.

Objectives

- Name the 6 essential elements of life
- Identify the two major categories of energy
- Define the two Laws of Thermodynamics
- Explain the process of photosynthesis

Key Terms

- Carbohydrates
- Chemical Energy
- Chlorophyll
- Energy
- Entropy
- First Law of Conservation of Energy
- Inorganic molecules
- Kinetic energy
- Lipids
- Natural organic compounds
- Nucleic acid
- Organic molecules
- Photosynthesis
- Potential Energy
- Proteins
- Second Law of Thermodynamics
- Synthetic organic compounds

Chapter Resource

Complementary *full color* illustrations, photos, charts, and graphs are available for Chapter 6 by following this URL: https://tinyurl.com/7vxzdaks This digital resource will enhance your understanding of the chapter concepts.

Life

What is life? Life can be defined by the combination of the following characteristics:

- Life is organized
- Life acquires materials
- Life acquires energy
- Life responds
- Life reproduces
- Life develops
- Life adapts

Common Elements and Compounds of Life

The most common and essential elements that make up life or living systems are:

- Carbon (C)
- Hydrogen (H)
- Nitrogen (N)
- Oxygen (O)
- Phosphorous (P)
- Sulfur (S)

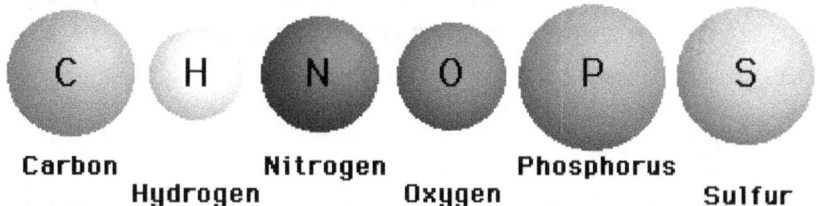

- Carbon
- Hydrogen
- Nitrogen
- Oxygen
- Phosphorus
- Sulfur

✓ Acronym CHNOPS makes it easy for you to remember them.
✓ These elements are the building blocks of all organic molecules that make up the tissues or plants, animals, and microbes.
✓ In living organisms, we find these elements bonded in large, complex molecules known as **proteins**, **carbohydrates** (sugars and starches), **lipids** (fatty substances), and **nucleic acids** (the natural macromolecules that function in the storage and transfer of genetic information).

- Some of these molecules may contain millions of atoms, and their potential diversity is infinite.

  o The diversity of living things is a reflection of the diversity of such molecules.

- The molecules that make up the tissues of living things are mainly constructed from carbon atoms bonded together into chains with hydrogen atoms attached and may include oxygen, nitrogen, phosphorus, and sulfur which are common building blocks of living organisms.

  o These carbon-hydrogen based molecules, which make up the tissues of living organisms, are called **organic molecules**.

  o **Inorganic molecules** or **compounds** have neither carbon-carbon nor carbon-hydrogen bonds, except Methane (CH_4), which is considered an organic compound.

  o Compounds of living organisms are referred to as **natural organic compounds** and the human-made ones as **synthetic organic compounds**.

- The lower atmosphere is a mixture of molecules of three important gases – Oxygen (O_2), Nitrogen (N_2), and Carbon dioxide (CO_2) – along with other trace gases, including variable amounts of polluting materials and water vapor.

  o Air is the source of carbon dioxide, oxygen, and nitrogen for all organisms.

  o Air is a mixture, meaning that there is no chemical bonding between the molecules involved.

- All other elements required by living organisms, as well as the 92 or so elements that are not required, are found in various rock and soil minerals.

- A mineral refers to any hard, crystalline, inorganic material of a given chemical composition. They are dense clusters of two or more kinds of atoms held together by electric charge.

Energy

- The universe is made up of matter and energy. **Energy** is the ability to move matter or to do work.

- Energy is commonly divided into two major categories: Kinetic and Potential.

    - **Kinetic energy** is energy in action or motion.
        - ✓ Light
        - ✓ Heat energy
        - ✓ Physical motion
        - ✓ Electrical current

    - **Potential energy** is energy in storage.
        - ✓ Rubber band
        - ✓ Firewood

    - Chemicals such as gasoline and other fuels (coal) release kinetic energy when ignited.

    - The potential energy contained in such chemicals and fuels is called **chemical energy**.

- Energy does not have mass or occupy space. Therefore, energy cannot be measured in units of weight volume.

    - It is measured in calorie, which is defined as the amount of heat required to raise the temperature of 1 gram (1 milliliter) of water 1 degree Celsius.

Energy Laws: Laws of Thermodynamics

- The **first law of conservation of energy** or the law of conservation of energy: Energy is neither created nor destroyed, but may be converted from one form to another.

- The **second law of thermodynamics**: in any energy conversion, you will end up with less usable energy than you started with.

    - Note that energy cannot be completely recycled because in every energy conversion some kinetic energy is lost as heat – that is energy cannot be changed from one form into another without a loss of usable energy.

    - A principle that underlies the loss of heat is the principle of increasing **entropy**.

- ✓ Entropy is the degree of disorder. The principle is that, without energy input, everything goes in one direction only, toward increasing entropy.

Matter and Energy Changes

- All organic molecules, which make up tissues of living organisms, contain high potential energy.

- When they burn, the heat and light of the flame are their potential energy being released as kinetic energy.

  - Inorganic molecules, such as carbon dioxide, water, or mineral compounds that occur in nature have very low potential energy. When they are burned, they do not produce energy, which is why they are used as fire extinguishers.

- The production of organic material from inorganic material involves a gain in potential energy. The breakdown of organic matter involves a release of energy.

- The relationship is demonstrated by the energy dynamics of ecosystems.

  - Producers (green plants) make high-potential energy organic molecules for their bodies from low potential energy raw materials in the environment, namely carbon dioxide, water, and a few dissolved compounds of nitrogen, phosphorus, and other elements.

  - This process is made possible by the sun light energy absorbed by **chlorophyll**, a process called **photosynthesis**.

    - ✓ During this process, sugar (glucose – stored chemical energy) is produced and oxygen gas is released as a by-product.

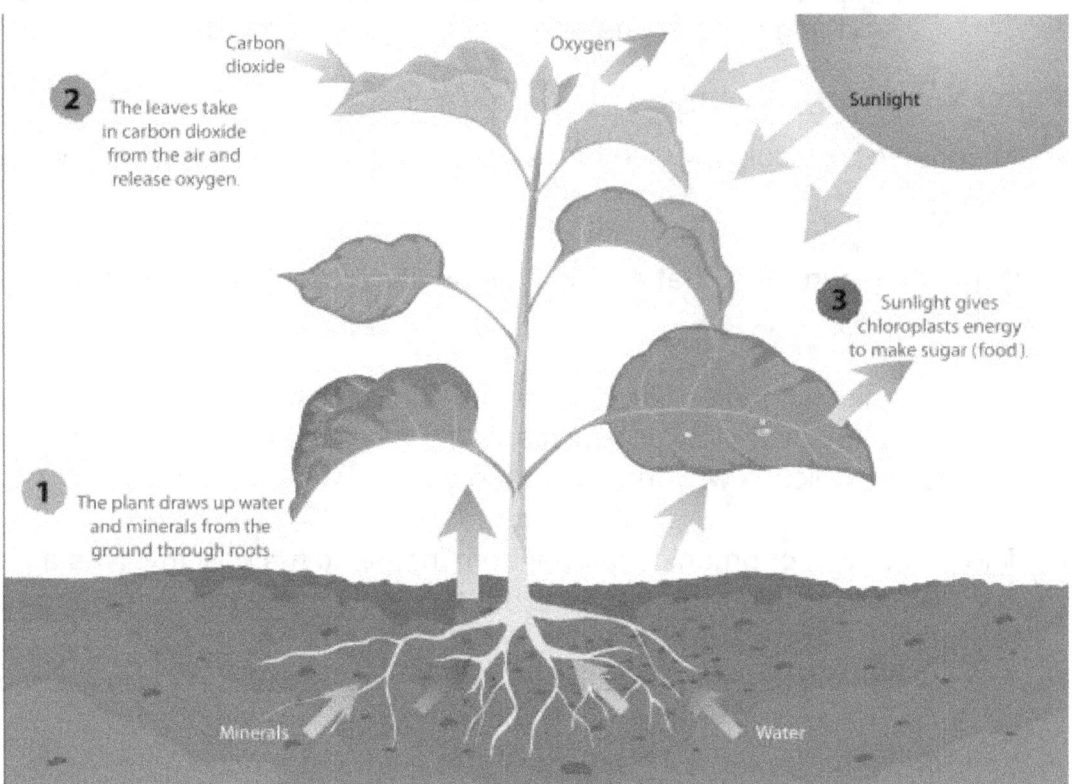

Additional Resources

Internet sites represent a vast resource of information. Using one of the search engines on the Internet such as Google or DuckDuckGo, find more information by searching for words, phrases, or websites: Always use caution when searching for information on the Internet. Follow guidelines to ensure the accuracy and reliability of the information you find.

Search words: Laws of Thermodynamics, kinetic energy, potential energy, chemical energy, organic molecules, and inorganic molecules.

Websites:
The Law of Conservation of Energy https://www.youtube.com/watch?v=_8EEnMwkmZk
Chemistry of Life www.khanacademy.org

Self-Assessment

1. List the essential elements that make up life.

2. What are the three important gases in the lower atmosphere?

3. List three forms each of kinetic energy and potential energy.

4. Describe the principle that underlies the loss of heat.

7 Photosynthesis

Major Concept

Photosynthesis uses radiant energy (light) from the sun to combine carbon dioxide (CO_2) and water (H_2O) to produce oxygen (O_2) and as glucose.

Objectives

- Describe the process of photosynthesis
- Recognize the chemical equation of photosynthesis
- Explain the role of glucose production
- Identify the waste production of photosynthesis

Key Terms

- Cellular Respiration
- Photosynthesis

Chapter Resource

Complementary *full color* illustrations, photos, charts, and graphs are available for Chapter 7 by following this URL: https://tinyurl.com/7vxzdaks This digital resource will enhance your understanding of the chapter concepts.

Photosynthesis

- **Photosynthesis** is a complex process that takes place in green plants.

- Radiant energy from the sun is used to combine carbon dioxide (CO_2) and water (H_2O) to produce oxygen (O_2) and carbohydrates (such as glucose ($C_6H_{12}O_6$)) and other nutrient molecules. The process is briefly explained below.
- The kinetic energy of light is absorbed by chlorophyll in the cells of the plant and used to remove the hydrogen atoms from water (H_2O) molecules.

  o The hydrogen atoms are transferred to carbon atoms coming from carbon dioxide as the carbons are joined in a chain to begin forming a glucose molecule.

$$6CO_2 + 6H_2O \xrightarrow[\text{Enzymes}]{\text{Energy}} C_6H_{12}O_6 + 6O_2$$

6 carbon dioxide 6 water   glucose 12 oxygen

- o After the removal of hydrogen from water, the oxygen atoms that remain combine with each other to form oxygen gas, which is released into the air.

- o 6 carbon atoms, 12 hydrogen atoms, and 6 oxygen atoms are assembled to form a molecule of glucose ($C_6H_{12}O_6$).

- The assembling of one molecule of glucose requires 6 molecules of carbon dioxide to provide the 6 carbon atoms and 6 molecules of water to provide 12 hydrogen atoms.

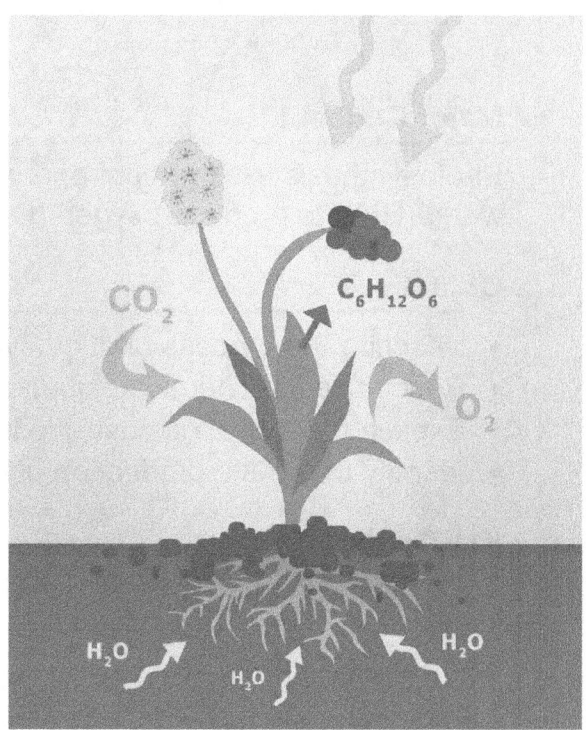

- Among these molecules of carbon dioxide and water are 18 oxygen atoms, but only 6 are needed.

  - o The extra oxygen atoms are given off as molecules of oxygen gas (O_2), 6 molecules for every molecule of glucose formed.

    - ✓ The oxygen, which is vital for the respiration of animals, is a waste product of photosynthesis.

- The glucose produced plays three roles in the plant.

  1. It is a raw material (either alone or in combination with nitrogen, phosphorus, sulfur, and other mineral nutrients) used for making all the other organic molecules (proteins, carbohydrates, etc.) that make up the stems, roots, leaves, flowers, and fruits of the plant.

  2. Synthesis of all organic molecules in the plant itself, absorption of minerals from the soil, and other plant functions require energy.

     - ✓ This energy is provided when the plant breaks down a portion of the glucose to release its stored energy in the process called **cellular respiration**.

  3. A part of the glucose produced may be stored for future use. Glucose is usually converted into starch, as in potatoes, or to oils, as in seeds for storage purposes.

Additional Resources

Internet sites represent a vast resource of information. Using one of the search engines on the Internet such as Google or DuckDuckGo, find more information by searching for words, phrases, or websites: Always use caution when searching for information on the Internet. Follow guidelines to ensure the accuracy and reliability of the information you find.

Search words: photosynthesis, kinetic energy of light, cellular respiration

Websites: Relationship between Photosynthesis and Cellular Respiration
https://www.youtube.com/watch?v=rXzN89I4_Yk

Self-Assessment

1. Write the chemical equation for photosynthesis and briefly describe the process.

2. List the three roles that glucose plays in photosynthesis.

3. What is the waste product of photosynthesis and why is it so important?

4. Describe the process of cellular respiration.

8 Cellular Respiration

Major Concept

All living organisms need energy to perform various activities, including movement, growth, maintenance, and repair of the body. This energy comes from cellular respiration.

Objectives

- Describe the process of cellular respiration
- List the two types of respiration
- Identify the nutrients needed for a balanced diet

Key Terms

- Aerobic respiration
- Anaerobic respiration
- Cell respiration
- Cellulose
- Fermentation
- Malnutrition
- Oxidation
- Perspiration
- Transpiration

Chapter Resource

Complementary *full color* illustrations, photos, charts, and graphs are available for Chapter 8 by following this URL: https://tinyurl.com/7vxzdaks This digital resource will enhance your understanding of the chapter concepts.

Cellular Respiration

- All living organisms need energy to perform various activities, including movement, growth, maintenance, and repair of the body. The energy comes from the breakdown of organic molecules of food.

- Starches, fats, and proteins are digested in the stomach and/or intestine in the case of consumers such as humans, which means they are broken into simpler molecules i.e., starches into sugar (glucose).

- The simple molecules are absorbed from the intestine into the bloodstream and transported to individual cells of the body.

    o Inside each cell, organic molecules (glucose) may be broken down through a process called **cell respiration**, where energy is released to perform the activities of the organism.

        ✓ The breakdown of simple organic molecules requires oxygen.
        ✓ The products of the process are carbon dioxide and water.
        ✓ This process is the reverse of photosynthesis. The process is expressed by the following formula.

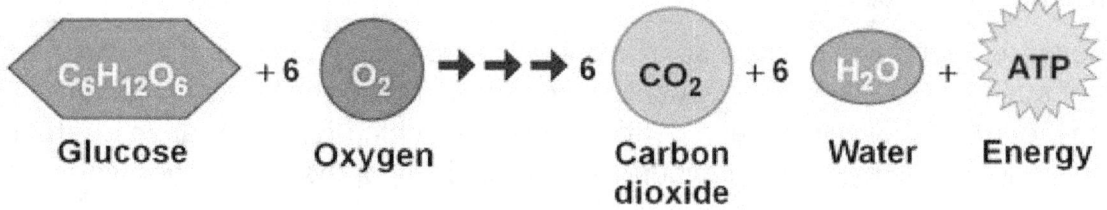

Cell respiration (an energy-releasing process)

- Carbon dioxide is released as waste product into the air and used by plants during photosynthesis.

- Water, which is another byproduct, may serve any of the body's needs for water or may be released through **perspiration** in the case of humans and **transpiration** in the case of plants.

- The chemical process that generally involves the breakdown of organic material, through combining with oxygen, is called **oxidation**.

    o Both burning and cellular respiration are examples of oxidation.
    o In both cases, organic matter is combined with oxygen and broken down to carbon dioxide and water.

- There are two types of respiration: **aerobic** that requires oxygen as stated above, and **anaerobic** that is oxygen free, which is carried out by certain bacteria and yeast.

    o Methane or alcohol may be the byproducts of this process.
    o If such organisms carry out partial breakdowns of organic molecules in the absence of oxygen, the process is called **fermentation**.

- For organisms to perform normal bodily functions, they need a balanced diet, which should consist of the right kind and amount of nutrients.

CELLULOSE

- The nutrients include carbohydrates, proteins, vitamins, fats and oils, minerals, and roughage (fiber – consisting mostly of cellulose).

    o **Cellulose** is an organic macromolecule that is the prime constituent of plant cell walls.
    o Cellulose is also the major molecule in wood, wood products, and cotton.

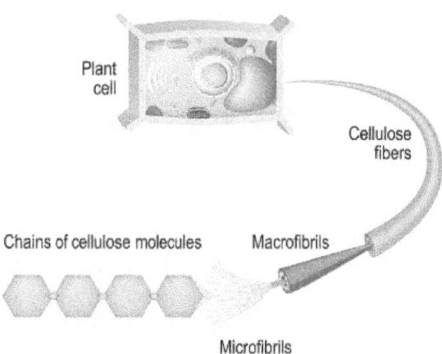

- If any one or more of the specific nutrients stated is absent from the diet a condition referred to as **malnutrition** develops.

Additional Resources

Internet sites represent a vast resource of information. Using one of the search engines on the Internet such as Google or DuckDuckGo, find more information by searching for words, phrases, or websites: Always use caution when searching for information on the Internet. Follow guidelines to ensure the accuracy and reliability of the information you find.

Search words: cellular respiration, transpiration, oxidation, aerobic, anaerobic, fermentation

Self-Assessment

1. Provide two examples of oxidation.

2. What happens during the process of fermentation?

3. Describe how water is released during respiration for humans and plants.

4. Cellulose is the major molecule in what products?

9 An Ecosystem and How it Works

Major Concept

An ecosystem is a community of different species interacting with one another and with the chemical and physical factors in the environment

Objectives

- List the biotic components of an ecosystem
- Identify two types of biomes
- List the five taxonomy kingdoms
- Name the three categories of organisms in a biotic structure
- Explain feeding and non-feeding relationships
- Indicate five terrestrial and five aquatic abiotic factors in the environment

Key Terms

- Abiotic
- Aquatic biome
- Aquatic life zone
- Biodiversity
- Biomass
- Biomes
- Biotic
- Biotic structure
- Chlorophyll
- Community
- Competitive relationship
- Consumers (heterotrophs)
- Decomposers
- Detritus
- Detritus feeders
- Ecosystem
- Ecotone
- Ectoparasite
- Endoparasite
- Fauna
- Flora
- Food chain
- Food web
- Habitat
- Herbivore
- Herbivory
- High orders of consumer/carnivores
- Host
- Intolerant
- Law of Limiting Factors
- Limit of tolerance
- Limiting factor
- Mutualistic relationship
- Niche
- Omnivores
- Optimum
- Parasite
- Photosynthesis
- Population
- Predation
- Predator
- Prey
- Primary consumers
- Producers (autotrophs)
- Range of tolerance
- Secondary consumers/carnivore
- Species
- Stressful
- Structure
- Symbiosis
- Synergism
- Taxonomy
- Terrestrial biome
- Trophic level
- Zones of stress

Chapter Resource

Complementary *full color* illustrations, photos, charts, and graphs are available for Chapter 9 by following this URL: https://tinyurl.com/7vxzdaks This digital resource will enhance your understanding of the chapter concepts.

What is an Ecosystem?

- Life can exist in the context of an ecosystem, which can be either a terrestrial system or an aquatic system.
- An **ecosystem** is a community of different species interacting with one another and with the chemical and physical factors in the environment.

    o For example, a population of plants, animals, and microorganisms occupying an area together, they produce, transfer, circulate, and accumulate energy and materials through their activities. The activities include biological processes such as:

    - ✓ Photosynthesis
    - ✓ Respiration
    - ✓ Decomposition
    - ✓ Excretion
    - ✓ Growth
    - ✓ Movement
    - ✓ Herbivory
    - ✓ Predation

- The living organisms in an ecosystem are collectively called a **community**.

    o A community can be explained as the population of all species living and interacting in an area at a particular time.

- Ecosystems can be of any size or shape.

    o Examples of large ecosystems are:

    - ✓ Fresh water
    - ✓ Oceans
    - ✓ Deserts
    - ✓ Grasslands
    - ✓ Deciduous forests
    - ✓ Tropical rain forests
    - ✓ Tundra

    o Examples of small ecosystems are:

    - ✓ A pond
    - ✓ A rotting log
    - ✓ A grassland-forest ecotone

    o Sometimes we refer to this as **biota** or **biotic** community (all the populations of different plants, animals, and microbes occupying a given area).

    o The biotic components of an ecosystem are collectively referred to as **fauna** (animal life) and **flora** (plant life).

- Biologists also like to use the expression **biological diversity** or the term **biodiversity** to refer to the diversity of life in all its forms and levels of organization in an ecosystem.

- Animals, plants, and microorganisms are the three major forms.
- Genes, species, communities, and ecosystems/biomes are among the many levels of organization of diversity.

Biomes

- **Biomes** can be a terrestrial system or an aquatic system. The terrestrial biome is the land system. The aquatic system is sometimes referred to as an **aquatic life zone**.

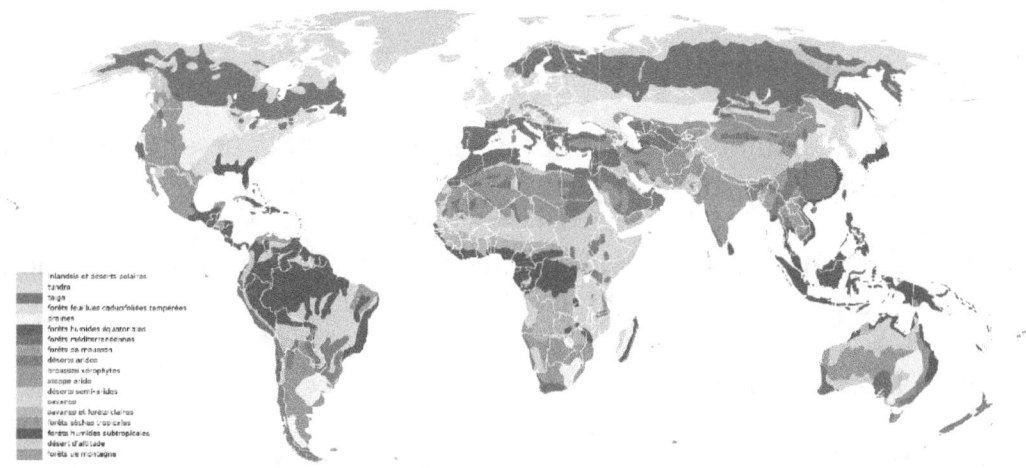

- The **terrestrial biomes** are regions inhabited by certain types of life or it is the grouping of all the ecosystems of a similar type present on biosphere (planet earth).

    o For example: tropical forest, grassland, desert, etc.

- An **aquatic biome** or an aquatic life zone, the watery parts of the biosphere include freshwater life zones (such as lakes, rivers, and streams) and ocean or marine life zones (such as coral reefs, coastal estuaries, and the deep ocean).

- Any terrestrial biome or aquatic life zone has two major categories of facts namely **biotic** factors – the living components (plants, animals and microorganisms sometimes referred to as biota) and **abiotic** factors – non-living components.

    o Examples of abiotic factors are:

    - ✓ Moisture
    - ✓ Temperature
    - ✓ Light
    - ✓ Solar energy
    - ✓ Nutrients
    - ✓ Water
    - ✓ Air
    - ✓ Wind
    - ✓ pH
    - ✓ Soil type
    - ✓ Salinity
    - ✓ Fire
    - ✓ Rainfall

    o In a natural system, due to abiotic factors, one biome (terrestrial or aquatic) tends to merge into one another.

- The gradual change of a tropical forest into grassland or one aquatic life zone merging into one of the terrestrial biomes.

- These transitional zones or regions between two biomes contain some of the species and characteristics of the two adjacent biomes.

- These transitional zones are called **ecotones**, the transitional region or zone in which one type of an ecosystem tends to merge with another ecosystem.

Taxonomy

- Because plants, animals and microorganisms have life, they are called living organisms.

- In order to understand how living organisms present in an environment interact, we must be able to classify them into groups and identify them by names.

- Living organisms can be broadly grouped into the following five Kingdoms (some texts say six) based on characteristics they exhibit.

1. Kingdom Monera (Monerans)

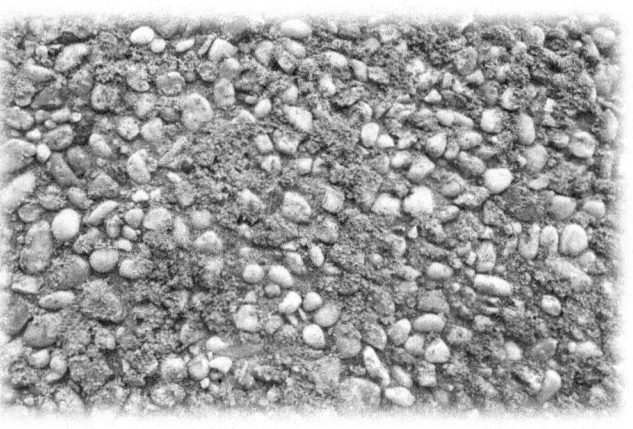

- Prokaryote, mostly unicellular with some colonial and multi-cellular forms
- Spherical, rod-like, and spiral
- Most with cell walls
- Lack a compact nucleus
- Genome is circular DNA molecule
- Lack intracellular membranous organelles
- Great metabolic diversity
- Asexual reproduction
- Microscopic

Examples:

- Cyanobacteria
    - ✓ Algal stromatolite
    - ✓ Oscillatoria

- Bacteria
    - ✓ *Aquaspirillum serpens*
    - ✓ *Rhizobium meliloti*
    - ✓ *Staphylococcus sp.*
    - ✓ *Echerichia coli*
    - ✓ *Lactobacillus acidophilus*
    - ✓ *Mycobacterium tuberculosis*
    - ✓ *Streptomyces griseus*

2. Kingdom Protista (Protists)

- Mostly unicellular eukaryotes, some multi-cellular
- Cellular structure simple to complex
- Autotropic nutrition, heterotrophic nutrition, or both
- Asexual reproduction common
- Most forms microscopic, but some reach 100 meters in length

Examples:

- *Globigerina sp.* (Foraminiferan Shells)
- *Volvox aureus* (Volvox)
- *Saproleginia sp.* (Water mold)
- *Dictyostelium discoideum* (Slime Mold)
- *Pfiesteria sp.* (Dinoflagellate)
- *Fucus sp.* (Rockweed)
- *Amoeba proteus* (Amoeba)
- *Pinnularia sp.* (Diatom)
- *Plasmodium vivax* (Malarial Parasite)

Rockweed

3. Kingdom Fungi (Fungi)

- Multicellular eukaryotes
- Chitinous cell walls
- Body of tubular hyphae forming a network called a mycelium
- Lack flagella
- Heterotrophic nutrition, including saprophytic, parasitic, and mutualistic forms
- Sexual and asexual reproduction
- Microscopic to several centimeters in diameter

Examples:

- *Morchella escullenta* (Morels)
- *Penicillium sp.* (Mold)
- *Amantina muscaria* (Fly Agaric)
- *Ustilago maydis* (Corn Smut)
- *Claviceps purpurea* (Ergot)
- *Cladonia cristatella* (British Soldier)
- *Saccharomyces cerevisiae* (Yeast)
- *Xanthoria sp. and Caloplaca sp.* (Lichens)
- *Rhizopus stolonnifer* (Bread Mold)

4. Kingdom Plantae (Plants)

- Multicellular eukaryotes
- Cellulose cell wall
- Photosynthetic autotrophs
- Reproduce sexually by forming gametes

Morel mushroom

- Alternating haploid (gametophyte) and diploid (sporophyte) generations
- Includes marine, freshwater, and terrestrial species ranging from microscopic to many meters in length.

- Two groups:

  1. Vascular Plants

  Examples:

  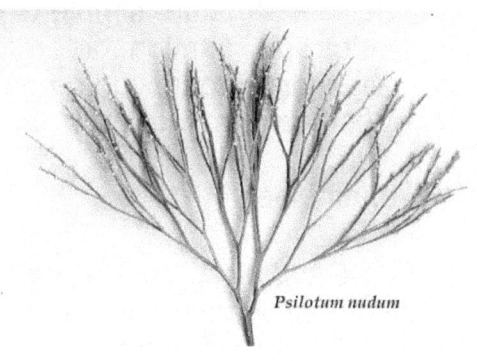

  - Monocot
    - ✓ *Zea mays* (Corn)
    - ✓ *Cypripedium acuale* (Orchid)
    - ✓ *Musa paradisiaca* (Banana Tree)
    - ✓ *Triticum aestivum aestivum* (Wheat)

  - Dicot
    - ✓ *Quercus sp.* (Oak)
    - ✓ *Helianthus sp.* (Sunflower)
    - ✓ *Glycine max* (Soybean)
    - ✓ *Rosa acicularis* (Wild Rose)

  - Other vascular plants
    - ✓ *Pseudotsuga menziesii* (Douglas Fir)
    - ✓ *Psilotum nudum* (Whisk Fern)
    - ✓ *Lycopodium* (Club Moss)
    - ✓ *Dryopteris arguta* (Wood Fern)

  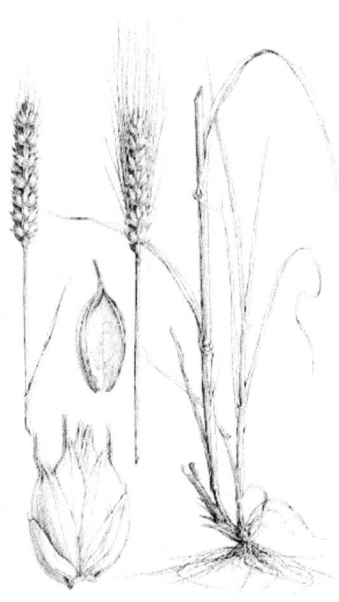

  2. Nonvascular Plants

  Examples:

  - ✓ *Polytricum* moss (Moss)
  - ✓ *Marchantia* (Liveworth)

5. Kingdom Animalia (Animals)

   - Multicellular eukaryotes
   - No cell wall
   - Hetrotropic
   - Sexual reproduction by forming gametes
   - Individuals develop from zygote formed by fusion of gametes

- Includes marine, freshwater, and terrestrial species ranging from microscopic to many meters in length

Two groups:

1. Vertebrates

    Examples:

    - ✓ *Alopex lagopus* (Arctic Fox)
    - ✓ *Terrapene carolina* (Eastern Box Tortoise)
    - ✓ *Entosphenus tridentatus* (West Coast Seas Lamprey)
    - ✓ *Rana pipiens* (Frog)
    - ✓ *Homo sapiens* (Human)
    - ✓ *Prionace glauca* (Blue Shark)
    - ✓ *Sula nebouxii* (Blue-footed Booby)
    - ✓ *Ursus arctos* (Brown Bear)
    - ✓ *Lepomis cyanellus* (Green Sunfish)

2. Invertebrates

    Examples:

    - ✓ *Scypha (Grantia)* (Sponge)
    - ✓ *Dugesia tigrina* (Brown Planaria)
    - ✓ *Asterias forbesi* (Starfish)
    - ✓ *Hirudo medicinalis* (Medicinal Leech)
    - ✓ *Octopus vulgaris* (Octopus)
    - ✓ *Astrangia danae* (Coral)
    - ✓ *Helix aspersa* (Snail)
    - ✓ *Cassiopeia xanmachana* (Upside-down Jellyfish)
    - ✓ *Rhabditis* (Roundworm)
    - ✓ *Pterourus (Papilio) glaucus* (Tiger Swallotail)
    - ✓ *Argiope aurantia* (Garden Weaver)
    - ✓ *Odontotaenius disjunctus* (Bessbug)
    - ✓ *Procambarus clarkii* (Crayfish)

- Viruses

    - Virus particles (virions) are submicroscopic
    - Obligate intracellular parasites of bacteria, protozoa, fungi, plants, animals
    - Consist of DNA or RNA fragments usually with protein coat (caspid), sometimes with an outer envelope of glycoprotein and lipid
    - No reproduction or metabolism outside host cell
    - Complex reproduction involving host cell ribosomes and separation of viral genome from its protein

Examples:

- ✓ Human Immunodeficiency Virus (HIV)
- ✓ Ebola Zaire
- ✓ Bacteriophage
- ✓ Advenovirus
- ✓ Tobacco Mosaic
- ✓ Pox
- ✓ SARS

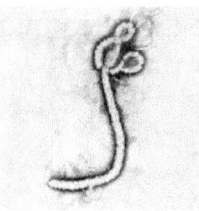
Ebola virus

- The grouping of organisms based on their natural characteristics is called **taxonomy**.

- The group of organisms able to breed among themselves to produce fertile offspring is referred to as **species**.

    o Species of one kind do not interbreed with species of other kinds.

    o All members of a particular species occupying a given area is referred to as a **population.**

- Examples are Bullhead catfish in an aquatic system, yellow corn in a farm and humans in a country.

- The study of an ecosystem will require an introduction to flora and fauna taxonomy at macro and micro levels.

The Structure of an Ecosystem

- Ecosystems can be large e.g., grassland biomes, or very small e.g., a rotting log.

- Structure of an ecosystem refers to parts and the way they fit together to make the whole.

    o There are two essential aspects to every ecosystem as previously stated: the biota or biotic community, and the abiotic environment.

- The way different categories of organisms fit together is referred to as **biotic structure**.

Biotic Structure

- All ecosystems have a similar biotic structure based on feeding relationships.

- These are divided into three categories of organisms:

    - ✓ Producers
    - ✓ Consumers
    - ✓ Detritus feeders (Decomposers)

Biotic Factors

Producers	Consumers	Decomposers

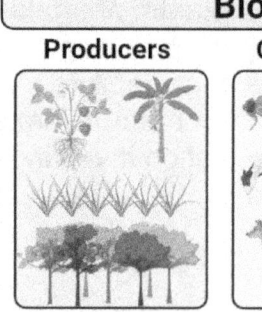  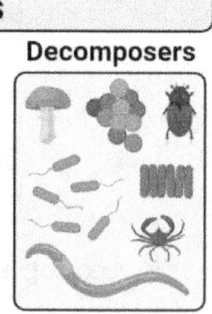

- **Producers** (**autotrophs**) are any organisms that can synthesize all their organic substances from inorganic nutrients or chemicals as a source of energy by the process of **photosynthesis**.

  ✓ Although green plants (use **chlorophyll** to capture energy from sunlight) are the principal autotrophs, photosynthetic bacteria (use purple pigment to absorb light energy to make food) and chemosynthetic bacteria (use high-energy inorganic chemicals such as hydrogen sulfide) are included in this category.

- **Consumers** (**heterotrophs**) are organisms that consume organic matter as a source of energy.

  ✓ They are divided into **primary consumers** (feed exclusively on plants), **omnivores** (feed on both plants and animals), **secondary consumers/carnivores** (feed on primary consumers), **high orders of consumers/carnivores** (feed on other carnivores) and **parasites** (organisms that attach themselves to another organism, called **host**, and feed on it for an extended period of time without killing it immediately).

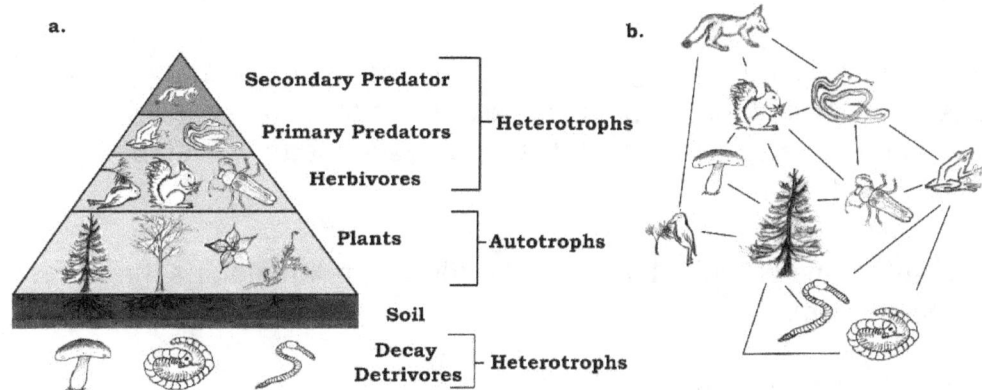

  ✓ Parasites are either **ectoparasites** (attach themselves outside the host e.g., ticks) or **endoparasites** (live inside the host e.g., worms, protozoa).
  ✓ The process of primarily feeding on green plants or plant products such as seeds or nuts is referred to as **herbivory**. The animal that feeds in this way is referred to as an **herbivore**.
  ✓ The process of one animal feeding on another animal is called **predation**. The animal that feeds in this way is referred to as a **predator** e.g., a lion feeds on a gazelle. The animal that is being fed on is referred to as **prey**.

- **Detritus feeders** (**detritivores**) are organisms such as termites, fungi, bacteria, earthworms, millipedes, crayfish, ants, and wood beetles that obtain their nutrients and energy mainly by feeding on dead organic matter. The dead matter is referred to as **detritus**.

- ✓ Two detritivores can be identified, primary detritivores (those that feed directly on detritus e.g., bacteria and fungi) and secondary detritivores (those that feed on primary detritivores e.g., protozoa, mites, insects, and worms).
- ✓ Primary detritivores are also referred to as **decomposers** (fungi and bacteria). They secrete digestive enzymes that cause the breakdown of wood, for example, converting them into simple sugars, which they absorb for their nourishment.

Feeding Relationships

- **Food chain** is the transfer of energy and material (biomass) through a series of organisms as each one is fed upon by the next.

- **Biomass** is the mass of biological material or total combined (net dry) weight of all the organism(s) in an area.

  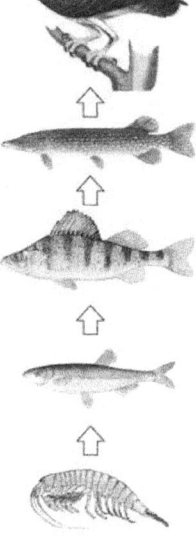

  o For example, if the biomass of producers in a grassland is 100,000 pounds per acre, the biomass of herbivores will be about 10,000 pounds and that of carnivores about 1,000 pounds. What happens to the other 90% at each level? Each level is referred to as a trophic level.

- **Trophic level** is the feeding levels with respect to the primary source of energy. Green plants (producers) are at the first trophic level, primary consumers at the second level and secondary consumers at the third level.

- **Food web** is the combination of all the feeding relationships that exist in an ecosystem.

Non-feeding Relationships

- **Mutualistic relationship** is a close relationship between two organisms in which both of them benefit from the relationship e.g., insects and flowers.

  o For example, a bee benefits by obtaining nectar from a flower whereas a plant benefits by being pollinated by the bee.

- Symbiotic relationship (**symbiosis**) is an association of two kinds of organisms. It does not specify a mutual benefit or harm.

  o Alternatively, it can be explained as the relationship that occurs when two different species live together in a unique way.

  o Their existence may be beneficial, neutral, or detrimental to one and/or the other species.

- **Competitive relationship** occurs when two or more organisms are making a common endeavor to gain one or more resource from the habitat they share. This usually occurs only when the resource is in short supply.

  - **Habitat** refers to the kind of place where a species is biologically adapted to live. A habitat may support a variety of animals. They may or may not compete in the habitat.

  - Each species as its own niche. A **niche** is the total way of life or role of the species in the community or ecosystem.

    - ✓ It includes all the physical, chemical, and biological conditions a species needs to live and reproduce in a community or an ecosystem, such as what it feeds on, where it feeds, when it feeds, where it finds shelter, and where it nests.

Abiotic Environmental Factors

- As stated earlier, abiotic factors refer to nonliving components of the environment.

  - Major terrestrial factors:

    - ✓ Rainfall
    - ✓ Temperature
    - ✓ Light
    - ✓ Wind
    - ✓ Chemical nutrients
    - ✓ pH (acidity)
    - ✓ Salinity (saltiness)
    - ✓ Fire

  - Major aquatic factors:

    - ✓ Salinity
    - ✓ Temperature
    - ✓ Chemical nutrients
    - ✓ Depth
    - ✓ Turbidity of water
    - ✓ Currents

    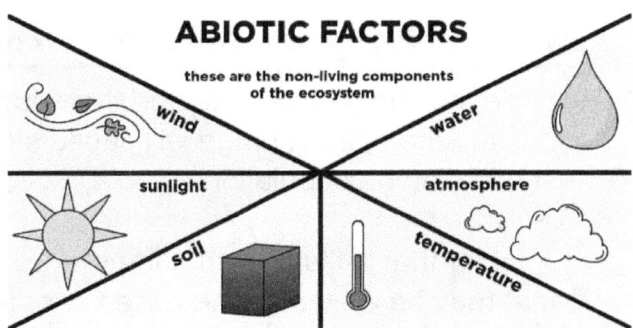

  - These factors determine which species may or may not occupy a given region or a particular area within a region.

- Environmental conditions may exist in different forms as experienced by different species. Conditions may be **optimum**, **stressful**, and **intolerant**.

- Different species thrive under different conditions such as wet, dry, warm bright sun, shady, salty, or freezing conditions. This principle applies to all living organisms.
- The condition or point at which the organisms respond best is called optimum.

- At higher or lower levels, the organisms do less well, and at further extremes, they may not be able to survive at all.
    - ✓ In this particular case, this often occurs over a range of several degrees. This range is called an **optimum range**.

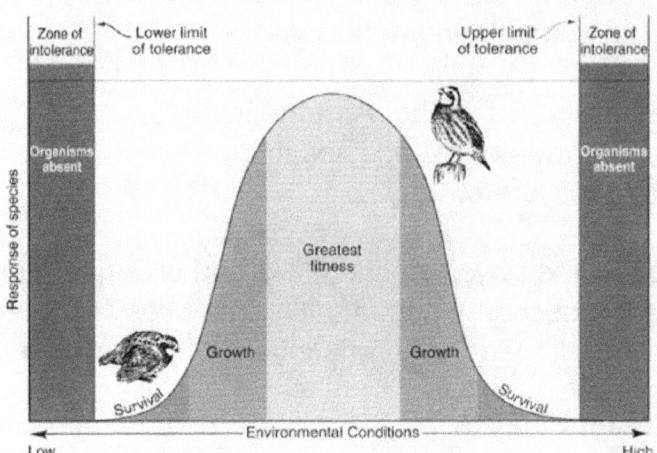

- The entire span that allows any growth at all is called the **range of tolerance**. The points at the high and low ends of the range of tolerance are called the **limit of tolerance**.
    - ✓ Between the optimal range and the high or low limit of tolerance, there are **zones of stress**.
    - ✓ Therefore, the population density of a species will be greatest where all conditions are optimal and vice versa.

Law of Limiting Factors (Leibig's Law of Minimums)

- For every single abiotic factor, there is an optimum limit of tolerance. Any factor being outside the optimal range will cause stress and limit the growth, reproduction, or even the survival of the population.

- A factor that limits growth and/or reproduction is called the **limiting factor** and the system that may be limited by the absence or minimum amount of any required factor is referred to as the **Law of Limiting Factors** or **Limiting Factors Principle**.

- When two or more factors interacting in a way that causes an effect much greater than one would anticipate from the effect of each of the two acting separately, we get what is referred to as **synergistic effects**, or **synergism**.

Different Regions Support Distinct Ecosystems

- Why do different regions support distinct ecosystems? This is attributed to the following:
    - Climate

- ✓ Temperature differences and precipitation availability and distribution – caused by altitude and/or latitude.

  o Microclimate and other abiotic factors
  o Local areas or sites may have temperature and moisture conditions that are significantly different from the overall climate of the region in which it is located.

  - ✓ E.g., a south-facing slope, which receives more direct sunlight will be relatively warmer than a north-facing slope.

  o Biotic factors

  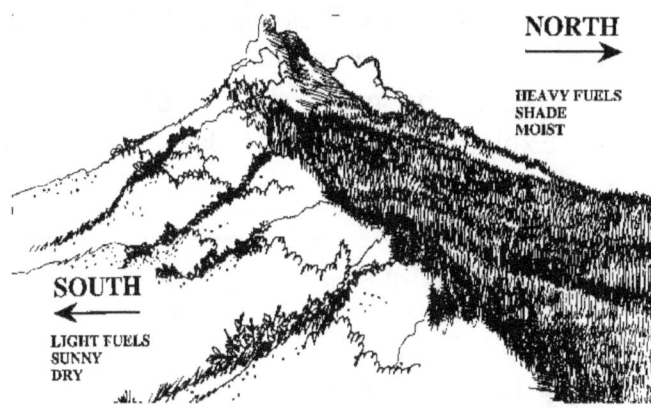

  - ✓ For example, under shady conditions of trees, the growth of grasses is limited. This biotic factor (shade from trees) limits the growth of grasses.

  o Physical barriers

  - ✓ Oceans, altitudes, latitudes, and mountain ranges have impacts on climates and hence distributions of organisms.

Additional Resources

Internet sites represent a vast resource of information. Using one of the search engines on the Internet such as Google or DuckDuckGo, find more information by searching for words, phrases, or websites: Always use caution when searching for information on the Internet. Follow guidelines to ensure the accuracy and reliability of the information you find.

Search words: ecosystem, biomes, biotic structure feeding and non-feeding relationships, Law of Limiting Factors

What are Ecosystems?
https://www.youtube.com/watch?v=9RPnsu_LkMY

Self-Assessment

1. List three examples of a large ecosystems and three examples of a small ecosystem.

2. Identify the differences between a terrestrial biome and an aquatic biome.

3. List the five taxonomy Kingdoms.

4. What are the three categories of organisms in a biotic structure?

5. Describe the difference between a mutualistic and competitive relationship.

10 The Principles of Ecosystem Functions

Major Concept

For sustainability, ecosystems dispose of wastes and replenish nutrients by recycling all elements.

Objectives

- Recognize the importance of nutrient cycling
- List the three nutrient cycles
- Explain the importance of solar energy
- Explain why land degradation should be prevented

Key Terms

- Nondepletable
- Nonpolluting
- Overgrazing
- Standing biomass

Chapter Resource

Complementary *full color* illustrations, photos, charts, and graphs are available for Chapter 10 by following this URL: https://tinyurl.com/7vxzdaks This digital resource will enhance your understanding of the chapter concepts.

Nutrient Cycling

- This is the first basic principle of ecosystem sustainability. Nutrient cycle is the repeated pathway or nutrients or elements from the environment through one or more organisms back to the environment.

- Nutrient cycles include the carbon cycle, the nitrogen cycle, the phosphorus cycle, etc.

- Nutrient cycle is important because:

  o It prevents wastes, which would cause problems when they accumulate.
  o It assures that the ecosystem will not run out of essential elements.

  > *These two important reasons lead to what is called as the first basic principle of ecosystem sustainability; for sustainability, ecosystems dispose of wastes and replenish nutrients by recycling all elements.*

Nutrient Cycles

- Carbon Cycle

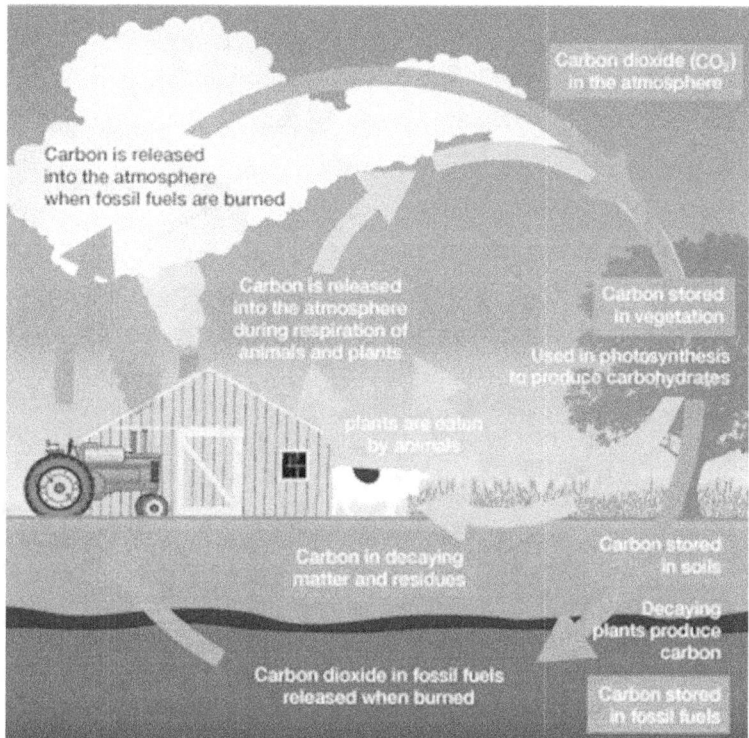

- Phosphorus Cycle

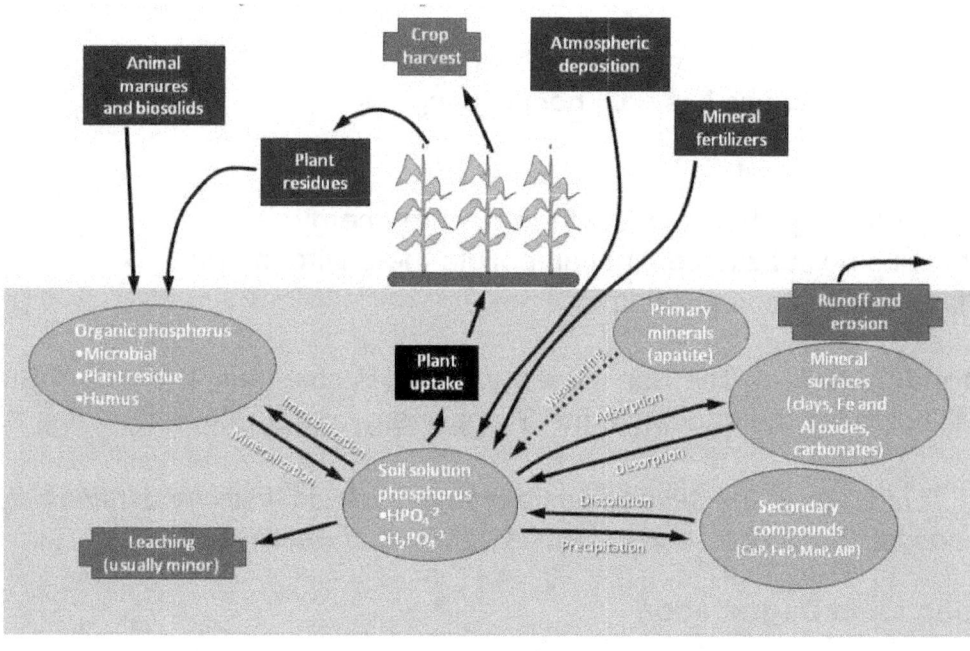

- Nitrogen Cycle

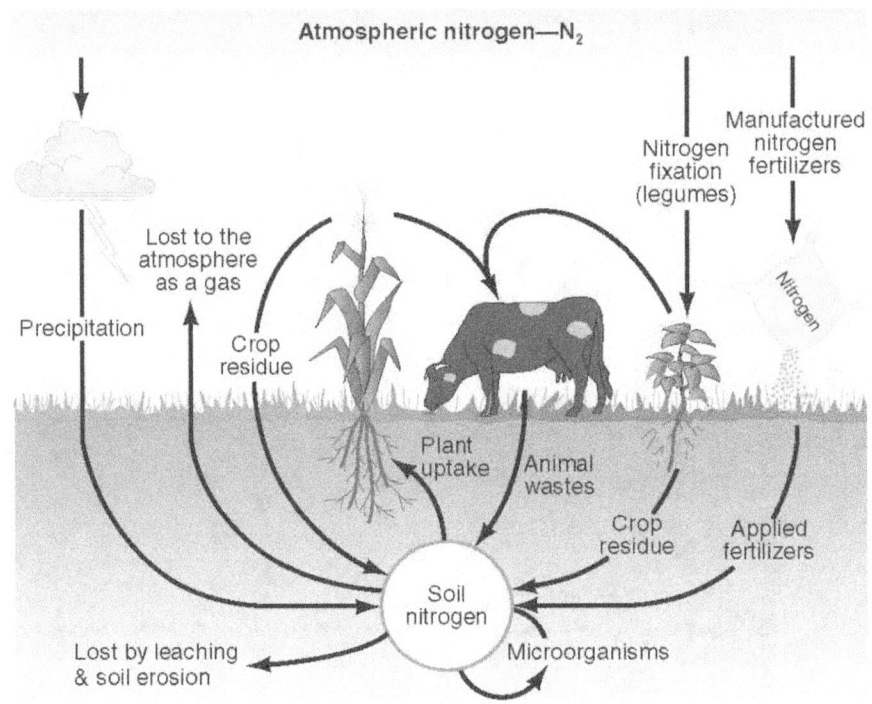

Running on Solar Energy

- This is the second basic principle of ecosystem sustainability. The initial source of energy major ecosystems (terrestrial and aquatic) is sunlight absorbed by green plants during photosynthesis.

- Sunlight as a source of energy is fundamental to sustainability because it is both nonpolluting and nondepletable.

    o **Nonpolluting** - Sunlight is a form of pure energy. It contains no substance that can pollute the environment; therefore, it is nonpolluting.

    o **Nondepletable** - The Sun's energy output is constant. The use of Sun energy on Earth will not influence or deplete the Sun's energy output; therefore, it is nondepletable.

    > *This leads to the second basic principle of ecosystem sustainability: for sustainability, ecosystems use sunlight as their source of energy.*

Preventing Land Degradation

- Preventing land degradation, for example, preventing overgrazing, is the third basic principle of ecosystem sustainability.

- In a grazing situation, it is readily apparent that the animals eat the grass faster than the grass can grow.

- Eventually the grass will be destroyed, and all the animals will starve. This situation is known as **overgrazing**. The same holds true for predators and their prey.

    o A sustainable situation demands that, on average, consumption cannot exceed production.

    o For this to happen, large portions of the producer must remain intact to maintain that productivity.

    o This portion or population that is not consumed and which remain intact to assure continued production is called the **standing biomass**.

- In natural ecosystems, which are sustainable, we observe that consumers eat no more than a small proportion of the total biomass available – most is left as standing biomass.

    > *This situation gives us the third basic principle of ecosystem sustainability, that is, for sustainability, the size of consumer population is maintained so that overgrazing or other overuse does not occur.*

Implications for Humans

- Contrary to the first basic principle of ecosystem sustainability, our human system is constructed based on a one-directional flow of elements.

    o Example, the fertilizer-nutrient phosphate, which is mined from deposits, end up going into waterways with effluents from sewage treatment.

    o The same one-way-flow is seen in metals such as Aluminum (Al), Mercury (Mg), Lead (Pb) and Cadmium (Cd), which are byproducts from industries.

    o This leads to a pollution problem in our water systems.

- Contrary to the second basic principle of ecosystem sustainability, our human system is heavily dependent on fossil fuels. The byproducts from fossil fuels pollute our air.

- Contrary to the third basic principle of ecosystem sustainability, our human system overuse natural resources, which leads to loss of biodiversity, tropical forests, and other forms of land degradation.

Additional Resources

Internet sites represent a vast resource of information. Using one of the search engines on the Internet such as Google or DuckDuckGo, find more information by searching for words, phrases, or websites: Always use caution when searching for information on the Internet. Follow guidelines to ensure the accuracy and reliability of the information you find.

Search words: nutrient cycling, carbon cycle, phosphorus cycle, nitrogen cycle, nonpolluting, nondepletability, standing biomass, one-directional flow of elements

Self-Assessment

1. List the two reasons why nutrient cycling is important.

2. Why is sunlight so important as a source of energy?

3. List the three basic principles of ecosystem sustainability.

4 Describe the implications of land degradation.

11 Ecosystem Balance and Imbalance

Major Concept

For an ecosystem to remain stable over a long period of time, the population of each species in the ecosystem must remain more or less constant in size and geographic distribution.

Objectives

- Explain biotic potential vs environmental resistance
- Describe population explosions
- Compare populations at replacement level and at a dynamic balance
- Outline the mechanisms of population balance

Key Terms

- Biotic potential
- Carrying capacity
- Density-dependent
- Density-independent
- Dynamic balance
- Environmental resistance
- Exponential increase
- Mortality
- Population balance
- Population explosion
- Recruitment
- Replacement level
- Reproduction rate
- Reproductive strategies

Chapter Resource

Complementary *full color* illustrations, photos, charts, and graphs are available for Chapter 11 by following this URL: https://tinyurl.com/7vxzdaks This digital resource will enhance your understanding of the chapter concepts.

Ecosystem Balance

- Each species in an ecosystem exists as a population (a reproducing group).

- For an ecosystem to remain stable (retain the same mix of populations of different species) over a long period of time, the population of each species in the ecosystem must remain more or less constant in size and geographic distribution.

- On average, deaths must be equal to births.

  - The equilibrium between births and deaths is referred to as **population balance**.

Biotic Potential vs Environmental Resistance

- **Biotic potential** is the number of offspring (live births, eggs laid, or seeds/spores set in plants) that a species may produce under ideal conditions.

- Different species produce different number of offspring at one time. To have an effect on the size of subsequent generations, the offspring must survive and reproduce in turn.

- The process of survival through the early growth stages to become part of the breeding population is called **recruitment**.

  o As species have different biotic potential and recruitment, they use different **reproductive strategies** to ensure the continuity of the species.

  o Reproductive strategies are methods seen in nature to enhance the chance of subsequent generations.

    ✓ E.g., producing massive numbers of young but offering no care or protection vs. producing few young and caring for them.

  o In addition, population growth and geographical distribution are influenced by the ability of the animal to emigrate, or seeds to disperse to similar habitats in other regions; the ability to adapt to and invade new habitats; defense mechanisms; and resistance to adverse conditions and disease.

  o Considering all these factors for population growth together, every species has the capacity to increase its population when conditions are ideal.

  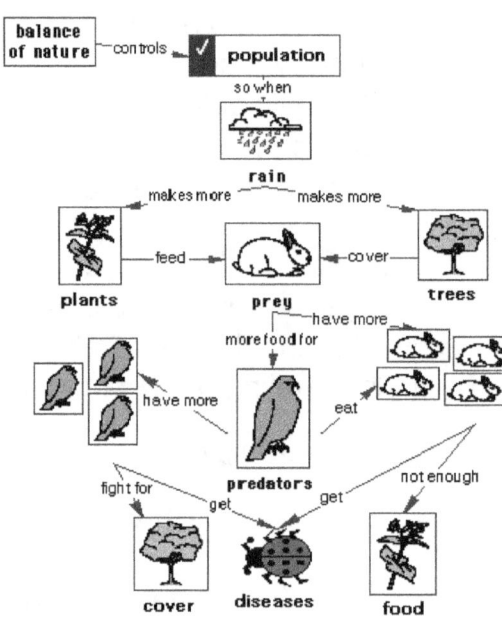

  o In the ideal condition, growth of population will be **exponential**.

    ✓ A pair of rabbits producing 20 offspring, 10 of which are female, may grow by a factor of 10 each generation: 10^1, 10^2, 10^3, 10^4, 10^5, etc.
    ✓ This type of increase is referred to as **exponential increase**. When this occurs in a population it is called **population explosion**.

- Population explosions are rare in natural ecosystems because a large number of both biotic and abiotic factors tend to decrease populations.

  o Among the biotic factors are:

    ✓ Predators
    ✓ Parasites
    ✓ Competitors
    ✓ Lack of food

  o Abiotic factors include:

    ✓ Unsuitable temperature
    ✓ Moisture
    ✓ Light
    ✓ Salinity
    ✓ pH
    ✓ Lack of nutrients

- All the biotic and abiotic factors that may limit population increase is referred to as **environmental resistance**.

- Because of the genetic endowment of species, generally, the reproductive rate of a species remains fairly constant.

- What varies substantially is recruitment, which effectively, is reduced by environmental resistance.

  - If the recruitment is at **replacement level** (the fertility rate that will just sustain a stable population) that is, just enough to replace these adults, then the size of the population will remain constant.

  - If the recruitment is not sufficient to replace losses in the breeding population, the size of the population will decline.

  - Therefore, because of interaction between biotic potential and environmental resistance, a population may remain stable or decreases.

    ✓ Usually, biotic potential remains constant; it is changes in environmental resistance that allows populations to increase or decrease.

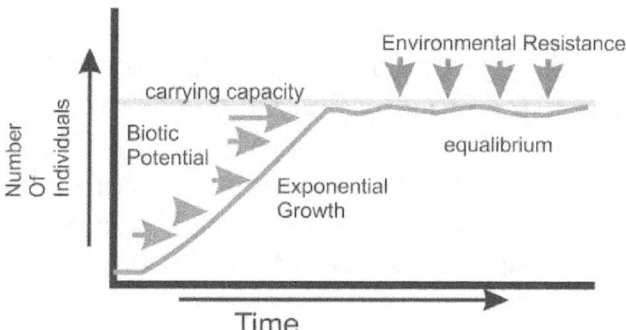

  - Population balance is said to be in **dynamic balance**, which implies that additions (births) and subtractions (deaths) are occurring continually, and the population may fluctuate around a median.

  - In addition, population survival depends on a certain minimum population base, which is referred to as the critical number - the minimum number of individuals of a given species that is required to maintain a healthy, viable population of species.

  - Usually, this number or more provide protection and support, and as a result, make breeding possible to its members.

  - If a population falls below the critical number, due to environmental and human factors, recovery of the population may be difficult.

- ✓ Species whose populations are declining rapidly because of human impacts are classified as threatened.
- ✓ If the population is near critical number, the species may be classified as endangered.
- ✓ When all the individuals of a particular species die and all the genes lost forever, this is referred to as extinction.

Mechanisms of Population Balance

- Predator-prey and host-parasite balances

    o Predator-prey relationship: this is a feeding relationship that exists between two kinds of animals. The predator is the animal feeding on the prey.

    o This relationship is vital in controlling populations of herbivores in most ecosystems.

    o All these depend on several factors including abundance of predators, type and abundance of prey, and numerous environmental factors.

    - ✓ For example, as a prey population is culled down to those healthy individuals that can escape attack, the predator population has no choice but to starve to a lower level.
    - ✓ At the time, the survivors of the prey population are the healthiest of the stock and can quickly reproduce the next generation. This is one of the mechanisms employed in a population balance of an ecosystem.

    o Host-parasite relationship: this feeding relationship is when an organism referred to as a parasite feeds on an organism called a host (i.e., supporting the feeder).

    - ✓ In terms of population balance, parasitic organisms act in the same way as large predators do.

    o In a food web, a population of any given organism is affected by a number of predators and parasites concurrently.

- Ecosystem out-of-balance (absence of balance)

    o A good example is when rabbits were introduced to Australia in 1895 with no natural enemy capable of controlling the rabbit population in the area of introduction.
    o Over time, the rabbit population exploded and devastated vast areas of rangeland by overgrazing.

- ✓ The destruction was remarkably damaging to both native wildlife and sheep ranching.
- ✓ It was brought under control, at least to some extent, by the introduction of a rabbit disease virus (myxomatosis) that provided a host-parasite balance.

- Territoriality

  o This is a behavioral characteristic exhibited by many animal species, especially birds and mammalian carnivore, to mark and defend an area (territory) against other members of the same species.

    ✓ Examples are birds singing in spring, and wolves, which would kill other wolves, dogs, and coyotes to keep their territory.
    ✓ As a result of territoriality, wolf numbers in the wild do not explode, unlike domestic dogs and cats.

- Plant-herbivore balance

  o Loss of this balance results in overgrazing, which can result in reduced carrying capacity.

    ✓ For example, too many elephants in parks in tropical and subtropical Africa, especially during drought, results in excessive loss of young Acacia spp. (trees), which lead to loss of tree reproduction/regeneration, which then lowers the carrying capacity for the elephants.

- **Carrying capacity**

  o This is defined as the maximum population of an animal that a given habitat will support without the habitat being degraded over the long term. Carrying capacity is a function of habitat or ecosystem.

    ✓ For example, a square mile of forest has a certain carrying capacity for deer or mice.

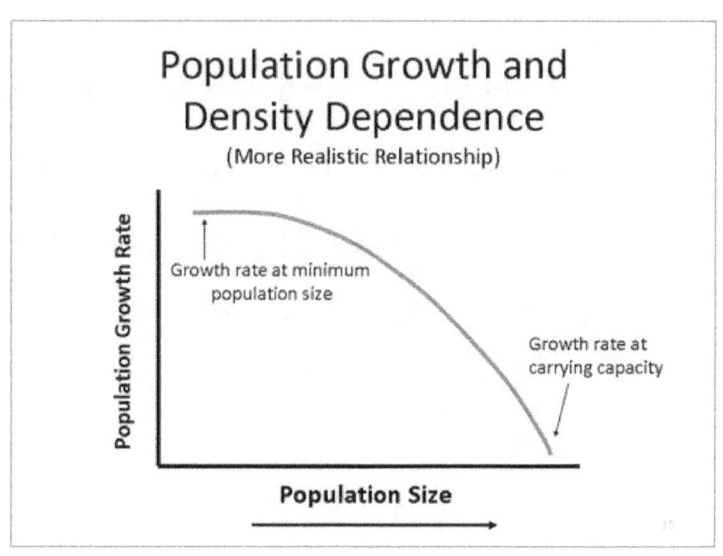

- Population growth

    o There are two major categories of population growth.

    o **Density-dependent** – this is a growth influenced by the density of the population.

        ✓ For example, if individuals of a species are introduced in a stable environment, the individuals will initially increase geometrically at a rate determined by the species biotic potential.
        ✓ However, as the density of the population increases, the environment becomes limiting as the habitat resources decrease. This leads to overuse of the best resources and the population growth rate decreases and the mortality increases.
        ✓ A point is reached when the reproduction rate is equal to mortality. At that point, the population stabilizes.
        ✓ Environmental resistance determines this point of stabilization.

    o **Density-independent** – this growth is not dependent on the density of the population, e.g., a population that exists in unlimited environment (no environmental resistance).

        ✓ As a result, the population increases at a rate determined by its biotic potential and may only be slowed down by severe environmental fluctuations such as drought or absence of cover.

Additional Resources

Internet sites represent a vast resource of information. Using one of the search engines on the Internet such as Google or DuckDuckGo, find more information by searching for words, phrases, or websites: Always use caution when searching for information on the Internet. Follow guidelines to ensure the accuracy and reliability of the information you find.

Search words: population dynamics, reproductive strategies, exponential increase, dynamic balance, territoriality, plant-herbivore balance

Self-Assessment

1. Provide examples of biotic and abiotic factors that tend to decrease populations.

2. Describe reproductive strategies for two different species.

3. List three organisms that engage in a host-parasite relationship.

4. Other than the rabbits in Australia, describe another situation where the ecosystem was out of balance.

12 Ecological Succession

Major Concept

Ecological succession is a change in a community following a disturbance or it is the process of gradual and orderly progression from one biotic community to another.

Objectives

- Compare primary succession and secondary succession
- Describe the factors that contribute to aquatic succession
- Identify factors of natural succession

Key Terms

- Aquatic succession
- Ecological succession
- Primary succession
- Secondary succession

Chapter Resource

Complementary *full color* illustrations, photos, charts, and graphs are available for Chapter 12 by following this URL: https://tinyurl.com/7vxzdaks This digital resource will enhance your understanding of the chapter concepts.

Ecological Succession

- **Ecological succession** is a change in a community following a disturbance or it is the process of gradual and orderly progression from one biotic community to another.

- Two types of ecological successions.

  - **Primary successions** occur in areas where there is no soil formation such as following a volcanic eruption or a glacial retreat.

    ✓ It can also be explained as the gradual establishment through a series of stages in an area that has not been occupied before.

  - **Secondary succession** begins in an area where soil is present, as when cultivated fields like cornfields return to a natural state.

    ✓ The first to begin secondary succession are called pioneer species, that is plants that are invaders of disturbed areas and then progresses through a series of stages that become stable for a long period of time.
    ✓ This may be referred to as the climax of the ecosystem. For example, from abandoned agricultural fields in the eastern United States back to deciduous forests.

  - Note that the major difference between primary and secondary succession is that secondary succession starts with preexisting soil substrate. Therefore, the early, prolonged stages of soil building are bypassed.

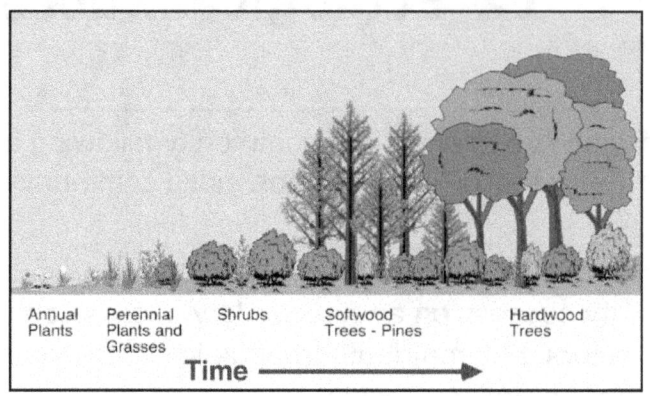

Climax Ecosystem

- This is the last stage in ecological succession – an ecosystem in which populations of all organisms are in balance with each other and with existing abiotic factors.

- **Aquatic succession** is seen in a lake or pond, which is gradually filled and taken over by the surface surrounding terrestrial ecosystem.

    o This occurs due to soil deposition at the bottom of the lake or pond, gradually filling them up. The aquatic vegetation produces detritus that also contributes to the filling process.

    o As the process of buildup occurs, terrestrial species can advance, and aquatic species move farther in the lake.

    o Finally, the lake disappears and is replaced by a terrestrial ecosystem.

Succession and Biodiversity

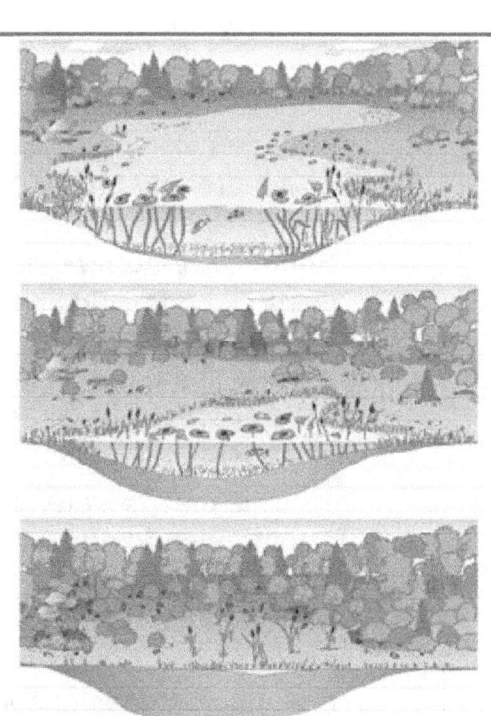

- The natural succession depends on maintaining the biodiversity of the surrounding area.

- Fire has been part of ecosystems since time immemorial and therefore regarded as a natural factor.

    o Fire has produced negative and positive effects on the biotic community and abiotic factors of ecosystems and therefore contributes to succession in many ways.

- Succession depends on preserving biodiversity, and succession underlies the ability of an ecosystem to recover from damage or overuse.

- Humans should increase their participation in restoring damaged areas of the ecosystems.

  > *Therefore, this leads to the fourth principle of ecosystem sustainability that is, for sustainability, biodiversity must be maintained.*

Additional Resources

Internet sites represent a vast resource of information. Using one of the search engines on the Internet such as Google or DuckDuckGo, find more information by searching for words, phrases, or websites: Always use caution when searching for information on the Internet. Follow guidelines to ensure the accuracy and reliability of the information you find.

Search words: climax of the ecosystem, natural succession

Self-Assessment

Humans should increase their participation in restoring damaged areas of the ecosystems. Write a reflective essay of about 250 words on any restoration projects in your community or any that you are aware of.

13 Ecosystem – Adapting to Changes

Major Concept

Adaptation is an ecological change or evolutionary change in structure or function allows or enhances an organism's ability to survive and reproduce in an environment.

Objectives

- Explain differential reproduction and biological evolution
- Describe selective breeding and natural selection
- Recognize the significance of Darwin's finches
- List the four traits that humans can inherent from their parents

Key Terms

- Adaptation
- Biological evolution
- Chromosomes
- Deoxyribonucleic acid (DNA)
- Differential reproduction
- Gene pool
- Genes
- Genetic variation
- Metabolism
- Natural selection
- Selective breeding
- Selective pressure

Chapter Resource

Complementary *full color* illustrations, photos, charts, and graphs are available for Chapter 13 by following this URL: https://tinyurl.com/7vxzdaks This digital resource will enhance your understanding of the chapter concepts.

Selection by the Environment

- Within a species population, certain individuals reproduce much more than others do. This difference in reproduction is called **differential reproduction**.

    o Differential reproduction will lead to a gradual modification in the gene pool as some genes become increasingly common in the population and other genes diminish in frequency of occurrence.

    o This change in gene pools of species over the course of generations is the basic nature of **biological evolution**.

    o Through the process of biological evolution, species are adapted to the environment in which they live.

- **Adaptation** is an ecological or evolutionary change in structure or function that produces better adjustment of an organism to its environment and hence enhances its ability to survive and reproduce.

- To understand this, we need to have some knowledge of genetics, which is the study of heredity and processes by which inherited characteristics are passed from one generation to the next.

- The cell nucleus contains **chromosomes**, which carries the **genes**, the unit of inherited material.

- The chromosomes consist of one very long strand of **Deoxyribonucleic acid (DNA)**.

- Genes are responsible in determining the characteristics of individuals of the same species, which results in variation among individuals. For example, eye color in humans.

    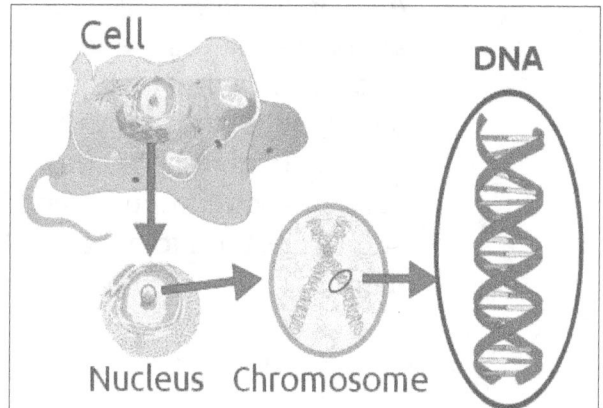

    o This range of variation that occurs among individuals of the same species is referred to as **genetic variation**.

    o The sum total of all the genes that exist at a particular time, among all the individuals of a species, is called **gene pool**.

Change through Selective Breeding

- Selective breeding refers to the breeding of individuals because they bear certain traits and exclusion from breeding with others.

    o For example, a breeder will select from an existing population of cattle those individuals that show the desired traits more than other members of the population. The selected individuals are then bred.

    o The offspring tend to be like the parents, but some offspring express the particular trait more than the parents do, and others express it less.

    o Those offspring that show the trait most are selected to be the parents of the next generation, whereas the other offspring are prevented from breeding.

    o Repeating this process of selection and breeding over many generations gradually yields the trait desired by the breeder.

Change through Natural Selection

- **Natural selection** is the process whereby the natural factors of environmental resistance tend to eliminate those members of a population that are least adapted to cope with those environmental factors and thus, in effect, select those best adapted for survival and reproduction.

- The various environmental resistance factors that can cause individuals with certain traits (which are not the norm for the population) to survive and reproduce more than the rest of the population is called **selective pressure**.

- The result is a transformation in the genetic makeup of the population.

  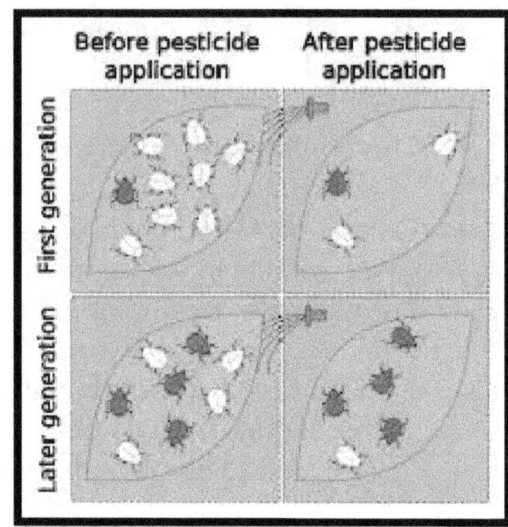

  ✓ Example: the presence of insecticide provides a selective pressure to increase pesticide resistance in the pest population.
  ✓ Also, in the presence of a predator, prey animals having traits (such as high speed or coloration that blends in with the background) that allow them to escape or protect them from their enemies tend to survive and reproduce, and those without such traits tend to become the predator's meal.

- Any individual with a gene that manifests itself as a hindrance will be eaten.

- Thus, predators may be seen as a selective pressure favoring the reproduction of genes that enhance the prey's ability to escape or protect itself and causing the elimination of any genes limiting those functions.

- In addition, the need for food is selective pressure acting on the predator, enhancing those characteristics benefiting predation – such as high speed, keen eyesight, sharp claws, or powerful jaws.

• Every factor of environmental resistance is a selective pressure resulting in the survival and reproduction of only those individuals with a genetic endowment enabling them to cope with their surroundings.

• In nature there is constant selection resulting in a modification of the species' gene pool toward features that enhance survival and reproduction within the existing biotic community and environment.

Adaptation to the Environment

• In natural selection, the main purpose for selection in every generation is survival and reproduction.

• Breeder's selection purpose is to obtain certain desired trait that could be beneficial to him.

• Under the selective pressures exerted by the environmental resistance, the gene pool of each population is modified such that the population becomes increasingly well adapted for survival and reproduction in the particular biotic community and environment in which it exists.

- All characteristics of organisms can be considered vital as follows:

    ✓ Adaptations for coping with climatic and other abiotic factors.
    ✓ Adaptations for obtaining food and water in the case of animals, and nutrients, energy, and water in the case of plants.
    ✓ Adaptations for escaping from or protecting against predation and for resistance to disease-causing or parasitic organisms.
    ✓ Adaptations for reproduction – finding or attracting mates in animal populations, for pollination and setting seed in plant populations.
    ✓ Adaptations for migration in the case of animals, and dispersal of seeds in the case of plants.

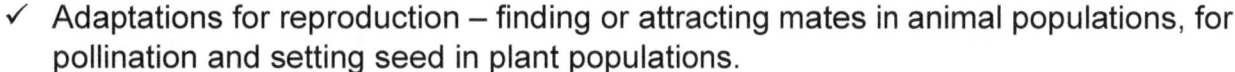

Selection of Traits and Genes

- Selections always take place at the level of the individual organism – that is, it is the individual organism that survives and reproduces, or ends up as somebody's meal.

- Yet, by differential reproduction weeding out certain individuals occurs while others reproduce.

    - This is because the genetic makeup of the individual often is the determining factor in its success or failure to survive and reproduce.

    - So, differential reproduction leads to a change in the gene pool of the population and hence modified traits.

- Traits are any physical or behavioral characteristics or talent an individual is born with. They are inherited from the parents.

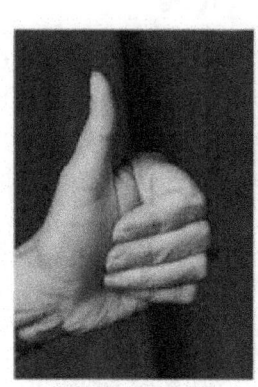

- Traits can be grouped as:
    - Physical Traits
        - Shape of the nose
        - Eye color
        - Body build
        - Lung capacity
        - Size of appendix
    - Metabolic Traits
        - Sensitivity to allergens
        - Digestive capacity
        - Tolerance to heat or cold
        - Resistance to disease
            - **Metabolism** is the sum of all the chemical reactions that occur in an organism.
    - Aptitude Traits
        - Running
        - Swimming
        - Jumping
        - Mathematics
        - Music
    - Behavioral Traits
        - Instinctive ways organisms act such as:
            - Spider spinning a web
            - Bird flying south for the winter
- Selective pressures acting on the gene pools of all species gradually lead to the development of an ecosystem that is in dynamic balance as long as conditions remain constant, which is not usually the case because environmental conditions always change over time.

Additional Resources

Internet sites represent a vast resource of information. Using one of the search engines on the Internet such as Google or DuckDuckGo, find more information by searching for words, phrases, or websites: Always use caution when searching for information on the Internet. Follow guidelines to ensure the accuracy and reliability of the information you find.

Search words: environmental adaptation, genetic mutation, natural selection

Self-Assessment

1. Provide an example of how an animal adapted to or adapts to changes in the ecosystem.

2. Describe the term selective pressure.

3. Provide three examples of adaptions to the environment.

Notes

14 Wildlife & Wildlife Management

Major Concept

Wildlife management involves manipulating wildlife populations and their habitats for their welfare and human benefit.

Objectives

- List the three approaches of the Integrated Wildlife Damage Management Program
- Recognize the purpose of wildlife management
- Explain the idea of a wildlife management plan
- Identify the inhabitants of a forest ecosystem

Key Terms

- Biodiversity
- Forest Ecosystem
- Homogeneous
- Wildlife

Chapter Resource

Complementary *full color* illustrations, photos, charts, and graphs are available for Chapter 14 by following this URL: https://tinyurl.com/7vxzdaks This digital resource will enhance your understanding of the chapter concepts.

Wildlife

- In modern scientific explanation, the terms wildlife, biological diversity (**biodiversity**) and **forest ecosystem** are similar, but they are not the same.

- **Wildlife** is a word of recent origin, and although often associated with game birds and mammals, its meaning has gradually changed over the years to include many organisms.

    o In broader terms, it is defined as all forms of life that are wild (organisms that live on their own without human help) and thus includes all wild animals, plants, and microorganisms.

    o Using this broader definition of wildlife avoids the use of phrases as destroying wildlife habitat and good for wildlife because forest habitat for one species will create good habitat for another.

    o From this perspective, one of the most basic objectives of wildlife management (skillful care of wildlife populations and their habitats) is to maintain or restore all a region's native wildlife populations, sometimes with a particular emphasis on endangered species.

Great Horned Owl
Credit: Robert Savannah, U. S. Fish and Wildlife Service

- Wildlife management involves manipulating wildlife populations and their habitats for their welfare and human benefit.

    o The wildlife management approach skillfully deals with game species on sustainable yield basis by:

        ✓ Using laws to regulate hunting and fishing
        ✓ Establishing harvest quotas
        ✓ Developing population management plans
        ✓ Improving wildlife habitat
        ✓ Using international treaties to protect migrating species

- The first step in wildlife management is to decide which species are to be managed in a particular area.

    o Ecologists and conservation biologists emphasize preserving biodiversity, wildlife conservationists are concerned about endangered species, birdwatchers want the greatest diversity of bird species, and hunters want large populations of game species.

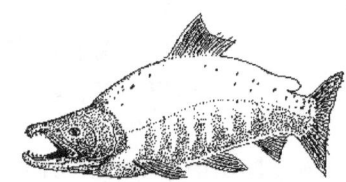

Credit: Robert Savannah, U. S. Fish and Wildlife Service

        ✓ In the USA and other developed countries, most wildlife management is devoted to producing surpluses of game animals and birds for hunters.

- Secondly, the wildlife manager must develop a management plan.

    o Ideally, this is based on the principles of:

        ✓ Ecological succession
        ✓ Wildlife population dynamics
        ✓ Understanding of the cover, food, water, space, predation herbivory and other habitat needs of each species to be managed
        ✓ The maximum sustained yield (MSY) of a population in which harvested individuals are not removed faster than they can be replaced through reproduction.

    o The manager must also consider:

        ✓ Number of potential hunters
        ✓ Their likely success rate
        ✓ Regulations for preventing excessive hunting

- Having a complete master management plan is a difficult undertaking because it is expensive and time consuming.

- Therefore, from personal experience, wildlife management plan involves educated guesswork and trial and error because most management plans are designed to be sensitive to political pressures from conflicting groups and budget constraints.

- This makes wildlife management plans an art rather than a science.

- Biological diversity (**biodiversity**) refers to the diversity of life in all its forms and levels of organization.

  - Animals, plants, and microorganisms are the three major forms.
  - Genes, species, communities, and biomes are among the many levels of organization.

- Managing for biological diversity is of critical importance because it is essential to the ecological well-being of the planet, upon which human welfare is ultimately dependent on.

- Every element of biological diversity has some economic or ecological value, although in many cases the economic value remains unrealized.

  - Ideally, natural resource managers will maintain all levels of biological diversity.
  - In practice, maintenance of species diversity will often be a fundamental goal because it is a relatively feasible and measurable goal.

- A broad approach to managing diversity is needed.

  - If you try to maximize diversity of specific species on every woodlot, all the woodlots would end up much the same. The landscape as a whole would be **homogeneous** (similar) and diversity on a larger, regional scale would be diminished.

  - Maintaining a diversity across the whole landscape is essential, rather than maximizing diversity on each piece of land or of specific species across a landscape.

- A **forest ecosystem** is a community of plants, animals, and microorganisms and the physical environment they inhabit, in which trees are observed to be the dominant life form.

- Trees are tall, long-lived plants because they are made of wood, a material of great strength and durability.

- o The height of trees, and thus of forests, produces considerable vertical stratification in terms of both the structure of trees and microclimate they create.

- The durability of wood allows large amounts of organic matter to accumulate.

  - o Together, vertical stratification and abundance of organic matter provide niches for a wealth of organisms, making forests especially diverse ecosystems.

  - o It takes time for trees to grow tall and for biomass to accumulate and thus succession, the process by which one group of species replaces another, makes forests even more diverse when viewed through a window of time.

  - o On a larger time scale, measured in thousands of years, forests are an ever-changing assemblage of species shaped by climate and other physical factors.

  - o In terms of spatial scale, we can think of forest ecosystems on the scale that makes them comparable to what foresters call a stand.

    - ✓ A stand is a group of trees that are reasonably similar in age, structure, and species composition, usually occupying at least 5 acres.

  - o However, a forest ecosystem is not a stand (stand refers to just the trees). Forest ecosystem refers to all the trees and their environment.

Integrated Wildlife Damage Management Program (IWDMP)

- The IWDMP consists of three approaches:

  1. Management of the resources being negatively affected

  2. Management of wildlife associated with the damage

  3. Physical separation of the two

- Resource management includes alteration of cultural practices, habitat modification, and alteration of human behavior.

- Management of wildlife includes behavior alteration through harassment, or scaring (non-lethal methods) and population manipulation through translocation or lethal removal.

- Physical separation may consist of fencing, netting, or other barriers.

Additional Resources

Internet sites represent a vast resource of information. Using one of the search engines on the Internet such as Google or DuckDuckGo, find more information by searching for words, phrases, or websites: Always use caution when searching for information on the Internet. Follow guidelines to ensure the accuracy and reliability of the information you find.

Search words: Wildlife management plans

Self-Assessment

1. Give an example of a wildlife management plan. What species needs to be controlled?

2. List twenty species that inhabit a forest ecosystem.

3. Describe how the wildlife management approach skillfully deals with game species on sustainable yield basis.

15 Pollution

Major Concept

Pollution is the addition of any material or heat energy in the amounts that cause undesired alterations in the physical, chemical, or biological characteristics of water, air, soil, or food that can adversely threaten the health, survival, or activities of humans or other living organisms.

Objectives

- Identify point and non-point sources of pollution
- List the different types of pollution
- Describe the damage pollutants can cause to the environment.
- Explain how pollution is created

Key Terms

- Non-biodegradable
- Non-point source
- Photochemical smog
- Point source
- Pollutant
- Pollution

Chapter Resource

Complementary *full color* illustrations, photos, charts, and graphs are available for Chapter 15 by following this URL: https://tinyurl.com/7vxzdaks This digital resource will enhance your understanding of the chapter concepts.

Pollution

- **Pollution** is the human-caused addition of any material or heat energy in the amounts that cause undesired alterations in the physical, chemical, or biological characteristics of water, air, soil, or food and therefore can adversely threaten the health, survival, or activities of humans or other living organisms.

- Any material that causes pollution is called a **pollutant**.

- Pollutants disrupt life support systems, cause damage to humans, domestic animals, and wildlife health, and produce unpleasant smells, tastes, sights, and sounds.

    o   Pollution is the byproduct of economic and social activities such as:

    - ✓ Producing crops
    - ✓ Creating comfortable homes
    - ✓ Providing energy
    - ✓ Transportation
    - ✓ Manufacturing products
    - ✓ Harnessing the atoms
    - ✓ Our basic biological functions (excreting wastes)

- Pollution problems have been increasing over the years due to the growing population and expanding per capita use of materials and energy.

- This increases the amount of byproducts that go into the environment, some of which are **non-biodegradable**.

- They resist attack and breakdown by detritus feeders and decomposers and consequently accumulate in the environment.

- Pollutants can enter the environment either naturally (i.e., from volcanic eruption) or through anthropogenic activities (i.e., from burning coal).

Sources of Pollutants

- **Point sources** are where pollutants come from single, identifiable sources such as a smokestack of a coal-burning power plant or paper mill, drainpipe of a factory, or an exhaust pip of an automobile.

- **Non-point sources** (difficult to identify the source) are where pollutants come from dispersed sources such as runoff of fertilizers and pesticides (from farmland, golf courses, suburban lawns, and gardens) into streams and lakes, and pesticides sprayed into the air or blown by the wind into the atmosphere.

- It is easier and cheaper to identify and control pollution point sources than from widely dispersed non-point sources.

Types of Pollution

- Water and land pollution

  - Sediments

    ✓ This is erosion from agriculture, mining sites, forest cutting, and construction sites.
    ✓ Sediments are soil particles, namely sand, silt, and clay, carried by flowing water, which then settles.

  - Nutrient oversupply

- ✓ This occurs in sewage treatment plants, and fertilizer runoff from lawns, gardens, and agricultural fields.

  o Toxic chemicals (harmful chemicals)

  - ✓ This occurs when leaching from disposal sites, direct discharges from certain industries, and spills take place.

  o Pesticide/Herbicides

  - ✓ This is a result of leaching from lawns, gardens, and agricultural fields.

  o Nuclear wastes

  - ✓ Nuclear power plants produce these wastes (wastes generated from nuclear reactor).

- Air pollution

  o Particulate matter

  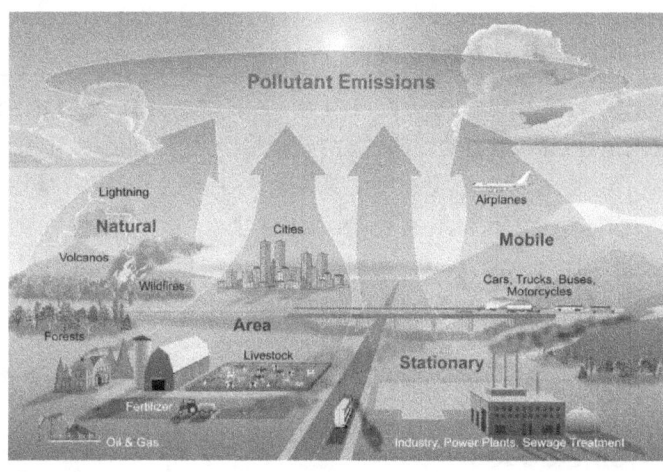

  - ✓ These are suspended particles smaller than 10 micrometers in diameter.
  - ✓ Such particles are readily inhaled directly into the lungs.
  - ✓ They are produced during refuse burning.

  o Acid forming compounds

  - ✓ They are compounds that form acidic solutions (less than pH 5.6) such as sulfur dioxide and oxides of nitrogen.
  - ✓ They are produced by coal-burning power plants and smelters.

  o Photochemical smog

  - ✓ Any chemical reaction activated by light is called photochemical reaction.
  - ✓ Air pollution known as **photochemical smog** is a mixture of primary and secondary pollutants formed under the influence of sunlight.
  - ✓ Photochemical smog is the brownish haze that frequently forms on otherwise clear, sunny days over large cities with a significant amount of automobile traffic.

- ✓ It results largely from sunlight-driven chemical reaction among nitrogen oxides and hydrocarbons, both which come primarily from auto exhausts.

  o Chlorofluorocarbons (*CFCs*)

    ✓ These are organic compounds made up of the atoms carbon, chlorine, and fluorine.
    ✓ An example is Freon-12 (CCl_2F_2) used as a refrigerant in refrigerators and air conditioners and in making plastics such as Styrofoam.
    ✓ Gaseous CFCs can deplete the ozone layer when they slowly rise into the stratosphere and there, chlorine atoms reaction with the ozone molecules, which have the potential of causing global warming and depletion of the ozone shield exposing living organisms to ultra-violet and other forms of dangerous sunray radiation.

Additional Resources

Internet sites represent a vast resource of information. Using one of the search engines on the Internet such as Google or DuckDuckGo, find more information by searching for words, phrases, or websites: Always use caution when searching for information on the Internet. Follow guidelines to ensure the accuracy and reliability of the information you find.
Search words: Air pollution, water pollution, toxic chemical pollutants, nuclear wastes, photochemical smog, particulate matter

Web sites: www.nps.gov, www.epa.gov

Self-Assessment

1. Provide five examples of point and non-point source pollutants.

2. What type of pollution do you feel needs to be addressed immediately? Why?

3. Give an example of nuclear waste and how it can adversely affect the environment.

4. What different types of anthropogenic activities contribute to worldwide pollution?

16 Water, Water Cycle & Water Purification/Treatment

Major Concept

All major terrestrial biota, ecosystems, and humans depend on fresh water – water that has a salt content of less than 0.01%.

Objectives

- Explain the three main loops of the water cycle
- List the different types of water
- Discuss the importance of water to all living organisms
- List the different methods of water purification

Key Terms

- Aquifer
- Condensation
- Evaporation
- Infiltration
- Percolation
- Precipitation
- Transpiration
- Water Cycle

Chapter Resource

Complementary *full color* illustrations, photos, charts, and graphs are available for Chapter 16 by following this URL: https://tinyurl.com/7vxzdaks This digital resource will enhance your understanding of the chapter concepts.

Definition and Importance of Water

- Water is a colorless, transparent, odorless, and tasteless liquid compound of oxygen and hydrogen. The chemical formula of water is H_2O.

- Water can be converted into steam by heat, and ice by cold temperature.

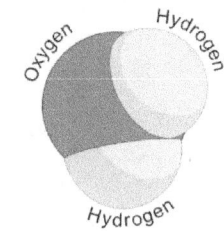

- Water is extremely important in maintaining vital processes to life and growth to all living organisms.

    o Water is used for:

    - ✓ Drinking
    - ✓ Washing
    - ✓ Bleaching
    - ✓ Dyeing
    - ✓ Cooling
    - ✓ Raising steam to drive engines or turbines to generate electricity
    - ✓ A solvent in industrial processes

Water Types

- All major terrestrial biota, ecosystems, and humans depend on fresh water – water that has a salt content of less than 0.01%.

- In addition to the term fresh water, we describe water in several ways. They are:

  - Water quantity – refers to the amount of water available to meet desired demand.
  - Water quality – means the degree to which water is pure enough to fulfill the requirements of various uses.
  - Saltwater – refers to water typical of oceans and seas that contains at least 3% salt.
  - Brackish water – a mixture of fresh and salt water, typically found where rivers enter the ocean.
  - Hard water – means water that contains minerals, especially calcium or magnesium, which cause soap to precipitate, producing a scum, curd, or scale in boilers.
  - Soft water – refers to water that is relatively free of those minerals that cause soap to precipitate causing scale buildup.
  - Polluted water – refers to water that contains one or more impurities making water unsuitable for a desired use.
  - Purified water – refers to water that pollutants were removed or rendered harmless.

Water Cycle

- The Earth's **water cycle**, also known as hydrologic cycle, is the movement of water from points of **evaporation** and **transpiration** to **condensation** and **precipitation**, and back to start the cycle again.

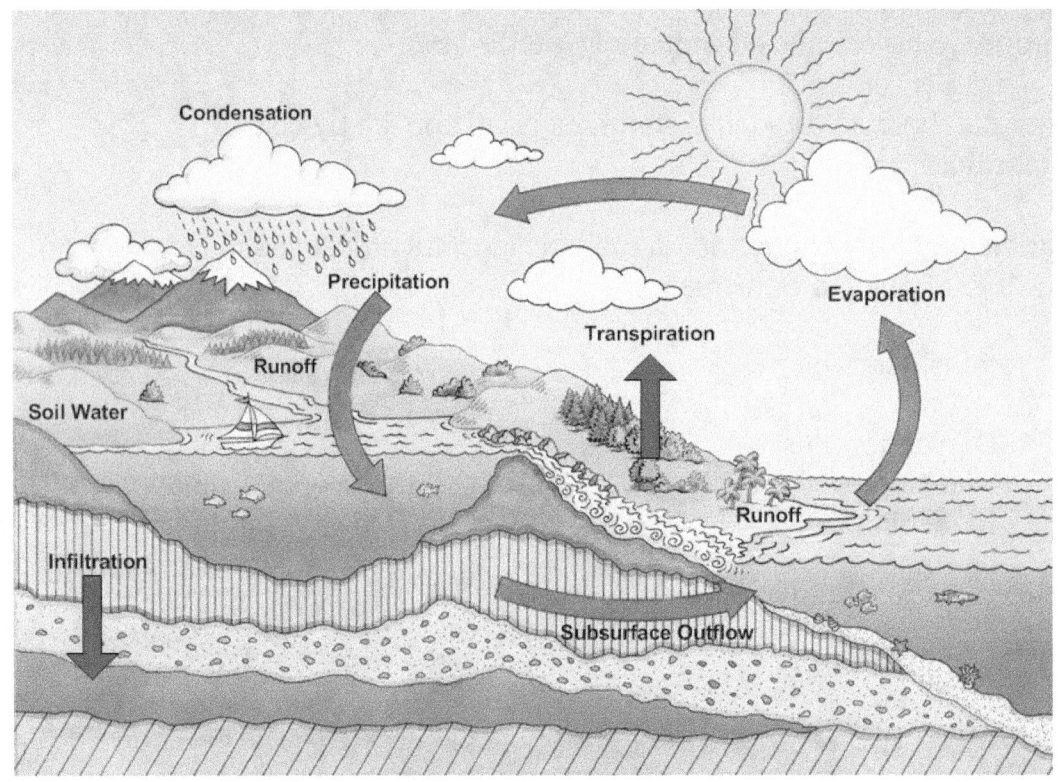

- There are three main loops in the cycle.

    1. The *surface runoff loop*, in which water runs across the ground surface and becomes part of the surface water system.

    2. The *evapotranspiration loop*, in which water **infiltrates**, is held in capillary water, and then returns to the atmosphere by way of evapotranspiration.

    3. The *groundwater loop*, in which water infiltrates, **percolates** down to join the ground water, and then moves through **aquifers** (underground layer of porous rock, sand, or other materials that allow the movement of water between layers of nonporous rock or clay). Finally, the water exits through springs, seeps, or wells where it rejoins the surface water to start the cycle again.

Methods of Water Purification

- Water purification is any method that will remove one or more impurities from polluted water.

- Several methods are used to purify water. Examples are:

    o Settling

    ✓ This refers to soil particles and other solid material carried by flowing water that may be removed by holding the water still and allowing the solids to settle. Clarified water is removed from the top.
    ✓ Settling may be aided by the addition of alum (aluminum sulfate). The +3 charge on the aluminum ions pulls clay and other particles, which are negatively charged, into clumps that settle more readily than the individual particles.

    o Filtration

    ✓ It is the passage of water through a porous material. Any material larger than the pores will be filtered out. A bed of sand is often used for this purpose.

    o Adsorption

    ✓ Certain materials bind and hold other materials on their surface. Passing water through an adsorbing material will remove certain pollutants.

    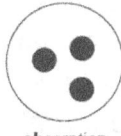 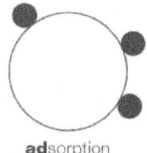

    ✓ Activated carbon is a material commonly used in this way to remove organic contaminants from water or air.

    o Biological Oxidation

- ✓ Organic material (detritus and organisms) is fed upon by detritus feeders and decomposers, broken down in cell respiration, and thus removed.
- ✓ Passage of water through systems supporting the growth of such organisms accomplishes removal of organic material (we will study this later).

  o Distillation

  - ✓ Distillation is the evaporation and condensation of water. All materials present in the water before the evaporation step remain behind in the holding tank and are therefore not present when the water vapor is condensed.

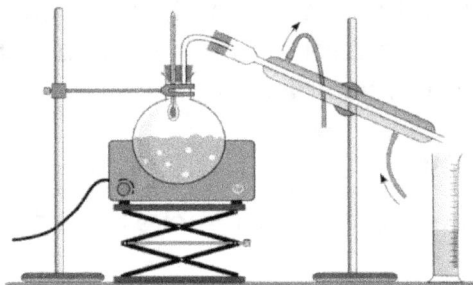

  o Disinfection

  - ✓ This refers to treatment of water with chlorine or other agents that kill disease-causing organisms.

Natural Water Cycle

- The natural water cycle includes all these purification methods except disinfection.

  o Sitting in lakes, ponds, or oceans, water is subject to settling.
  o As it percolates through soil or porous rock, it is filtered.
  o Soil and humus are also good chemical adsorbents.
  o As water flows down streams and rivers, detritus is removed by biological oxidation.
  o As water evaporates and condenses, it is distilled.

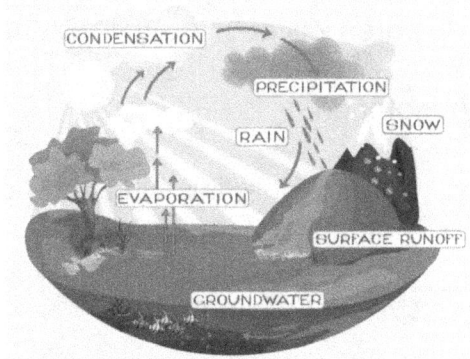

- Thus, numerous sources of fresh water might be safe to drink if it was not due to pollution problems caused by humans.
- One of the most serious water pollution problems is the threat of humans' health because of contamination with disease-causing organisms and parasites, which come from the excrements of humans and their domestic animals.

Water Treatment

- Water is often taken from a river, treated, used, and then returned.
- Water is also piped from a reservoir to a treatment plant. At the plant, several steps are taken to treat water.

  - Chlorine is added to kill bacteria.
  - Alum is added to coagulate organic particles.
  - The water is put into a settling basin for several hours to allow the coagulated particles to settle.
  - It is filtered through sand.
  - It is treated with lime to adjust pH and where there is fluoride deficiency, fluoride is added.
  - It is put into a storage water tower or reservoir for distribution.

Minidoka, Idaho Project – An Early Example of Water Control and Use

- The Minidoka Project is a series of public works by the U.S. Bureau of Reclamation to control the flow of the Snake River in Wyoming and Idaho, supplying irrigation water to farmlands in Idaho.

- One of the oldest Bureau of Reclamation projects in the United States, the project involves a series of dams and canals intended to store, regulate, and distribute the waters of the Snake, with electric power generation as a byproduct.

  - The water irrigates more than a million acres (4,000km^2) of otherwise arid land, producing much of Idaho's potato crop.

    ✓ Other crops include alfalfa, fruit, and sugar beets.

  - The primary irrigation district lies between Ashton in eastern Idaho and Bliss in the southwestern corner of the state.

  - Five main reservoirs collect water, distributing it through 1,600 miles (2,600 km) of canals and 4,000 miles (6,400 km) of lateral distribution ditches.

- History

    - Early studies of irrigation in southern Idaho began in 1889-90 by the U.S. Geological Survey. The data developed were made available to the Reclamation Service after the passage of the 1902 Reclamation Act.

    - The Minidoka Project was established in 1904, with construction of Minidoka Dam starting the same year.

    - Water could flow to the north bank of the Snake by gravity, but pumping was required for the south bank.

    - The project was designed to combine flood control and impoundment of spring runoff for use later in the growing season.

    - Jackson Lake and Palisades Reservoir are regulated to keep the flow at Heise, Idaho from exceeding 20,000 cubic feet per second ($570 m^3/s$).

    - The Minidoka Project contributed to the settlement of the Snake River Plain and river valley, converting semi-arid land to productive farmland. Population rose from a few thousand people in 1915 to more than 200,000 by the 1980s.

    - During the 1930s, the project used labor provided by the Civilian Conservation Corps (CCC) to construct canals.

    - Later, during World War II, 10,000 Japanese American internees were held at what is now Minidoka National Historic Site in Jerome County, Idaho. Internees worked on canal maintenance and provided agricultural labor.

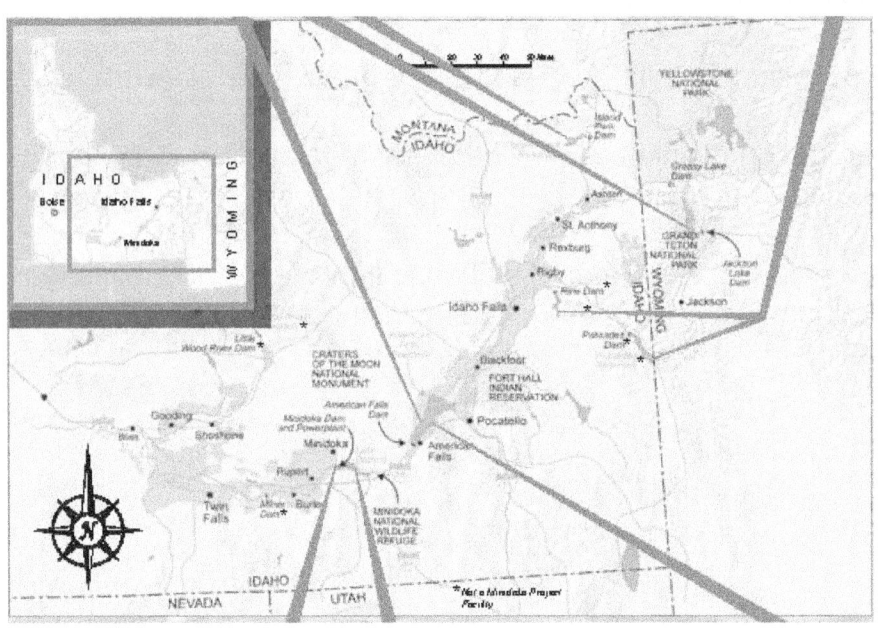

- Facilities

  o The project's dam and reservoirs comprise Jackson Lake Dam, Grassy Lake Dam, American Falls Reservoir, and Minidoka Dam (Lake Walcott).

  o Jackson Lake Dam in Grand Teton National Park, which raises the elevation of the natural glacial Jackson Lake by 30 feet (9.1 m), has a storage capacity of 847,000 acre-feet (1.045 km^3)

  o Grassy Lake Dam is in Wyoming in Bridger-Teton National Forest between Yellowstone National Park and Grand Teton National Park, directly adjacent to the south boundary of Yellowstone.

    ✓ Built between 1935 and 1939, Grassy Lake Dam is 118 feet (36 m) high with a capacity of 15,200 acre-feet (0.0187 km^3) on Grassy Creek.
    ✓ Grassy Creek's flow was not sufficient to supply the reservoir reliably, so water from Cascade Creek was diverted by a 14 feet (4.3 m) dam and diversion canal 0.7 miles (1.1 km) long.
    ✓ The lands around Grassy Lake and Island Park Reservoirs are administered by the U.S. Forest Service.

  o Island Park Dam on the Henrys Fork of the Snake River, 38 miles (61 km) north of Ashton, Idaho is 91 feet (28 m) and has a storage capacity of 135,000 acre-feet (0.167 km^3). It was built at the same time as Grassy Lake Dam.

    ✓ Along with the Grassy Lake Reservoir, the Cross Cut diversion dam, and the Cross Cut Canal, it forms the Upper Snake River Division of the Minidoka Project.
    ✓ The canal moves water from the Henrys Fork into the Teton River.

  o American Falls Reservoir on the main branch of the Snake River is the largest reservoir in the Minidoka Project, impounded by American Falls Dam.

    ✓ The original dam was built between 1925 and 1927 and was replaced between 1976 and 1978.
    ✓ Reservoir capacity is 1,672,600 acre-feet (2.0631 km^3).
    ✓ Construction of the dam and reservoir required the relocation of most of the town of American Falls, with many existing structures relocated.

  o Lake Walcott is impounded by Minidoka Dam on the Snake. Minidoka Dam started construction in 1904 to provide irrigation and power.

    ✓ The lake has a capacity of 210,200 acre-feet (0.2593 km^3).
    ✓ The power plant was one of the first to be installed in a Bureau of Reclamation project, providing power principally for pumping operations.
    ✓ Much of Lake Walcott is within Minidoka National Wildlife Refuge, providing habitat for waterfowl.
    ✓ The dam and power plant are listed on the National Register of Historic Places.

- The five main reservoirs are connected by a network of canals and pumping stations to regulate and distribute the water, supplying more than 1 million acres (4,000 km²).

- Associated Projects

  - Palisades Reservoir on the Snake River in Idaho and Wyoming is operated by the Bureau of Reclamation but is not part of the Minidoka Project.

    - ✓ Operated as the separate Palisades Project, the zoned earthfill dam is operated in coordination with the Minidoka and Michaud Flats projects, storing 1,401,000 acre-feet (1.728 km³) and generating up to 176 megawatts of power.
    - ✓ Palisades was built in the 1950s to address downstream water shortfalls.

  - The Michaud Flats project serves 11,240 acres (4,550 ha) with a pumping station on the left bank of the Snake, just below American Falls Dam, and 25 wells.

  - The Teton Basin Project on the upper Teton River in eastern Idaho included the Teton Dam and reservoir.

    - ✓ The dam failed on June 6, 1976 as the 288,250 acre-feet (0.35555 km³) reservoir was being filled, leading to extensive flooding in the Minidoka project area.

  - The Ririe Project in the Willow Creek drainage stores up to 100,500 acre-feet (0.1240 km³) of drainage from the Caribou Range.

Additional Resources

Internet sites represent a vast resource of information. Using one of the search engines on the Internet such as Google or DuckDuckGo, find more information by searching for words, phrases, or websites: Always use caution when searching for information on the Internet. Follow guidelines to ensure the accuracy and reliability of the information you find.

Search words: Water treatment, water cycle, Minidoka, Idaho Project

Website: US Bureau of Reclamation, Minidoka Project

Self-Assessment

1. Diagram the water cycle.
2. Outline the Minidoka water project.
3. Name and give examples of three types of fresh water.

4. How do these human activities introduce pollution into the water cycle?

Pollution by landfill
Industries
Sewage treatment
Chemicals

Waste drain
Agriculture farm chemicals
Aircraft
Automobiles

17 Eutrophication

Major Concept

Eutrophication, or the enrichment of water with plant nutrients such as nitrogen and phosphorus, can adversely affect lakes, reservoirs, impoundments, and coastal waters.

Objectives

- Describe eutrophication
- List the causes of eutrophication
- Identify the health problems in animals and humans due to eutrophication

Key Terms

- Eutrophication

Chapter Resource

Complementary *full color* illustrations, photos, charts, and graphs are available for Chapter 17 by following this URL: https://tinyurl.com/7vxzdaks This digital resource will enhance your understanding of the chapter concepts.

Eutrophication

- Natural ecosystems avoid pollution and resource depletion by recycling all elements – the first principle of ecosystem sustainability.

- Yet, humans have generally constructed systems based on a one-way flow.

- In particular, the pathway of fertilizer elements such as nitrogen, phosphorous, and potassium usually from crop soils through the food chain to humans and then into waterways with a discharge of sewage effluents. This results in pollution of our water systems.

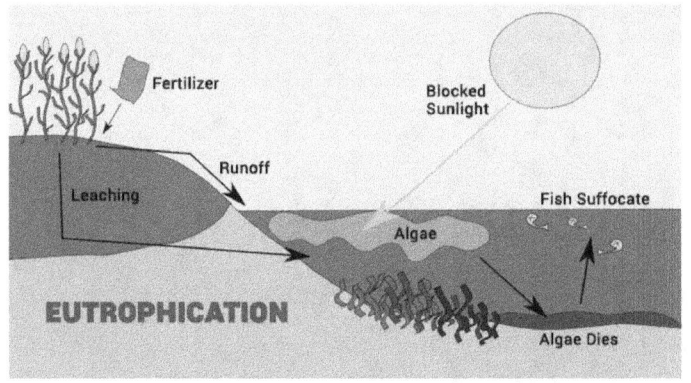

  o To mitigate such environmental pollution problems will involve the ability to:

   ✓ Identify the material(s) causing the pollution
   ✓ Identify the source of pollution
   ✓ Develop a plan for controlling the pollution
   ✓ Implement a plan for controlling the pollution

- **Eutrophication**, or the enrichment of water with plant nutrients, is a global phenomenon affecting many lakes, reservoirs, impoundments, and coastal waters.

- Although eutrophication can be a natural process over many centuries, accelerated eutrophication is associated with human activities worldwide.

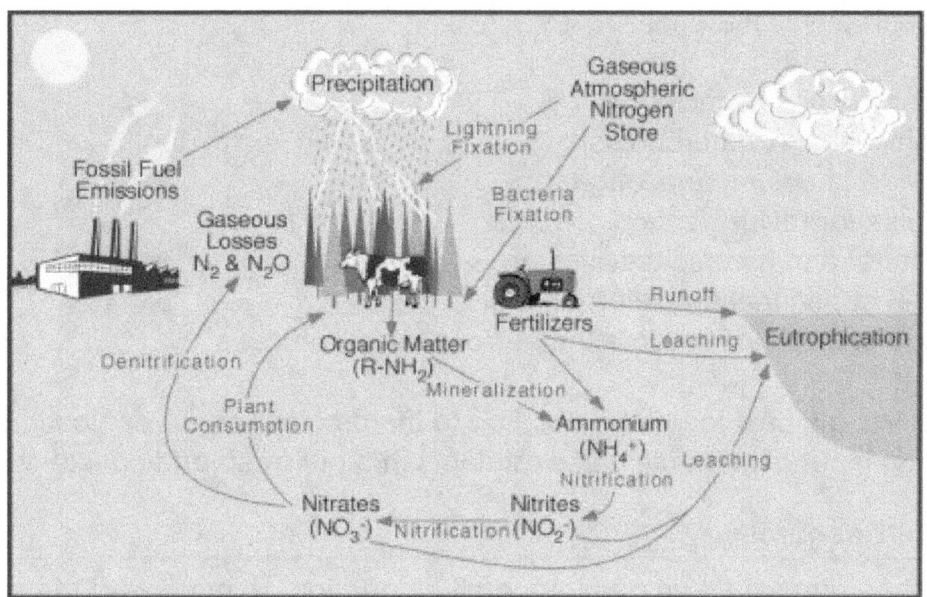

- Increased nutrient loads to fresh water and coastal waters arise from untreated urban wastes including detergent inputs, agricultural wastes, and fertilizer leachates, as well as industrial wastes containing nutrients (particularly from phosphate production process).

    o The dumping of sewage sludge in coastal areas provides a nutrient input to marine ecosystems.

    o Eutrophication seriously affects water quality although specific effects can be influenced by climatic, limnological, oceanographic, and biological factors.

    o Enrichment of waters, particularly with nitrogen and phosphorous, leads to enhanced plant growth whether in the form of microscopic algae or macrophyte agglomerations.

- Human and livestock health problems arise from consumption of water contaminated with toxin-producing algae.

    o Human health can also be seriously affected by consumption of shellfish infected with toxin-producing micro-algae and by increased development of disease vectors associated with macrophyte growth.

- Eutrophication process can also lead to ecological effects on the waterbody, particularly hypoxia, and consequential fish mortality when the enhanced alga and macrophytic materials decay.

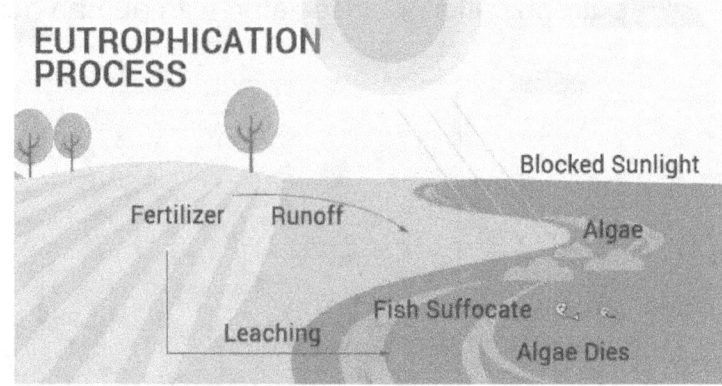

- With the increasing incidence of eutrophication in waterbodies worldwide, medium, and long-term strategies for rationalization of some agricultural and human activities are required at national and international levels to control and reverse the eutrophication process.

- Therefore, the best long-term solution to the problem of eutrophication is a program of reducing input of nutrients and sediments into our fresh and coastal water systems.

Additional Resources

Internet sites represent a vast resource of information. Using one of the search engines on the Internet such as Google or DuckDuckGo, find more information by searching for words, phrases, or websites: Always use caution when searching for information on the Internet. Follow guidelines to ensure the accuracy and reliability of the information you find.

Search words: hypoxia, estuaries, agricultural wastes, fertilizer leachates, sewage sludge, dead zones, algal blooms

Websites: www.oceanservice.noaa.gov/
https://www.wri.org/initiatives/eutrophication-and-hypoxia

Self-Assessment

1. Give examples of locations worldwide that are known as dead zones.

2. Identify possible causes of eutrophication in your own community. What can be done to prevent this?

3. Describe the implications of algal blooms and shellfish to animals and humans.

4. Discuss ways that are being used to reverse eutrophication.

18 Sewage Pollution

Major Concept

Sewage contains contaminants from wastewater, including household sewage and runoff (effluents) which need to be removed.

Objectives

- Explain sewage and why proper disposal and treatment is essential to human health and the environment
- List the four categories of pollutants in raw sewage
- Identify the three stages of sewage treatment
- Describe how storm water runoff affects sewage

Key Terms

- Sewage
- Primary treatment
- Secondary treatment
- Tertiary treatment

Chapter Resource

Complementary *full color* illustrations, photos, charts, and graphs are available for Chapter 18 by following this URL: https://tinyurl.com/7vxzdaks This digital resource will enhance your understanding of the chapter concepts.

Sewage

- **Sewage** (wastewater) is a major source of water contamination from pathogens (viruses, bacteria, protozoan, and worms) and from nutrients (especially phosphorus and nitrogen).

- The greatest hazard of domestic sewage waste is its potential for harboring disease-causing organisms.

- A special test used for monitoring sewage contamination in drinking water is fecal coliform test.

    - This is for the presence of *Escherichia coli (E. coli)*, the bacterium that normally inhabits the gut of humans and other mammals.

    - A positive test indicates sewage contamination and the potential presence of disease-causing microorganism carried by sewage.

- Usually, human populations continue to suffer from sewage-carried pathogens because humans live, work, and reproduce in situations of high population density.

    - One of the greatest risks of dense population is that, if a person becomes ill, disease can spread with great rapidity.

- Modern day good health is a result of clean water, sewage collection systems, effective sewage treatment, and sanitary food preparation.

Pollutants

- The wastewater stream entering a sewage treatment plant is generally 99% water and 1% polluting materials.

- Pollutants in raw sewage are divided into four categories:

    - Debris and Grit

        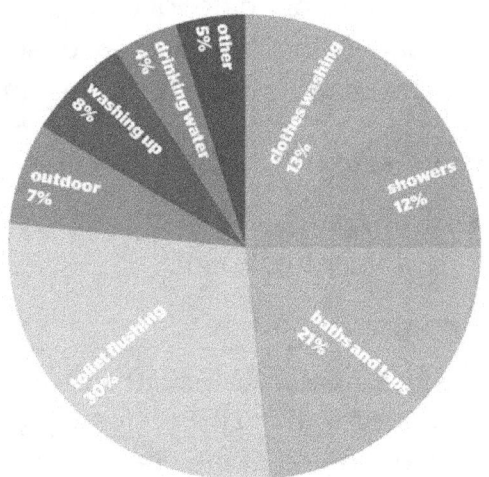

        - Debris includes rags, plastic bags, and other objects flushed down toilets or wasting through storm drains in places where they are still connected to sewers.
        - Grit is coarse sand and gravel, and it enters mainly through storm drains.

    - Particulate Organic Material

        - Particulate organic material includes visible particles of organic matter, originating from food wastes, home garbage disposal units, as well as fecal matter and bits of paper from toilets.
        - Also includes living bacteria and other microorganisms that have begun to digest the waste and possibly pathogenic organisms.
        - Particulate portion of organic material by definition consists of particles that will settle in still water.

    - Colloidal and Dissolved Organic Material

        - It originates from the same sources as particulate organic material; the difference is the particle size.
        - Whereas particulate organic material will settle in still water, colloidal particles are so fine that will not settle, at least not within any reasonable time period.

    - Dissolved Inorganic Material

        - It includes mainly the nitrogen, phosphorus, and other nutrients from excretory waste plus phosphate from detergents and water softeners.

- Other Contaminants
  - ✓ May include pesticides, heavy metals, and other toxic compounds discharged into sewers.

Sewage Treatment

- Sewage treatment is the process of removing contaminants from wastewater, including household sewage and runoff (effluents).
  - It includes physical, chemical, and biological processes to remove physical, chemical, and biological contaminants.
  - Its objective is to produce an environmentally safe fluid waste stream (or treated effluent) and a solid waste (or treated sludge) suitable for disposal or reuse (usually as farm fertilizer).

- Sewage collection and treatment is typically subject to local, state, and federal regulations and standards.

- Industrial sources of sewage often require specialized treatment processes.

- Sewage treatment generally involves three stages, called primary, secondary, and tertiary treatment.
  - **Primary treatment** consists of temporarily holding the sewage in a quiescent basin where heavy solids can settle to the bottom while oil, grease, and lighter solids float to the surface.
    - ✓ The settled and floating materials are removed, and the remaining liquid may be discharged or subjected to secondary treatment.
  - **Secondary treatment** removes dissolved and suspended biological matter.
    - ✓ Secondary treatment is typically performed by indigenous, water-borne microorganisms in a managed habitat.
    - ✓ Secondary treatment may require a separation process to remove the microorganisms from the treated water prior to discharge, or tertiary treatment.
  - **Tertiary treatment** is sometimes defined as anything more than primary and secondary treatment to allow release into a sensitive or fragile ecosystem (estuaries, low-flow rivers, coral reefs, etc.).
    - ✓ Treated water is sometimes disinfected chemically or physically (for example, by lagoons and microfiltration) prior to discharge into a stream, river, bay lagoon, or wetland.
    - ✓ It can also be used for the irrigation of a golf course, green way, or park.

✓ If it is sufficiently clean, it can also be used for groundwater recharge or agricultural purposes.

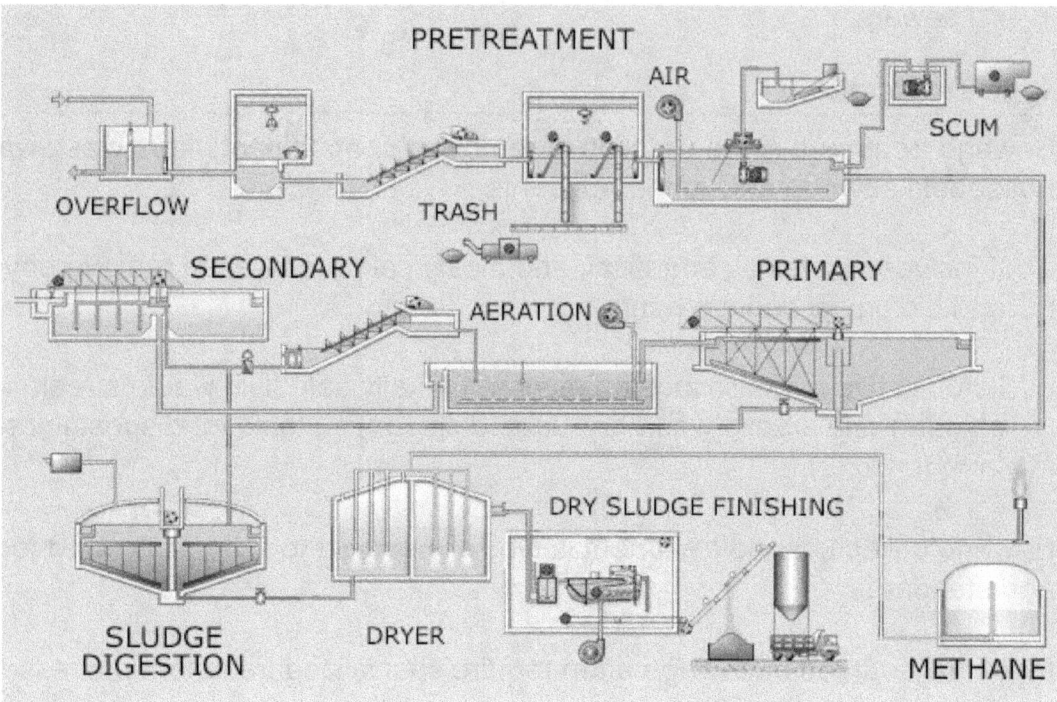

Origins

- Sewage is generated by residential, institutional, commercial, and industrial establishments.

    o It includes household waste liquid from toilets, baths, showers, kitchens, sinks, and so forth that is disposed of via sewers.

- In many areas, sewage also includes liquid waste from industry and commerce.

    o The separation and draining of household waste into greywater and black water is becoming more common in the developed world, with greywater being permitted to be used for watering plants or recycled for flushing toilets.

- Sewage may include storm water runoff. Sewerage systems capable of handling storm water are known as combined sewer systems.

    o This design was common when urban sewerage systems were first developed in the late 19th and early 20th centuries.
    o Combined sewers require much larger and more expensive treatment facilities than sanitary sewers.
    o Heavy volumes of storm runoff may overwhelm the sewage treatment system, causing a spill or overflow.

- - Sanitary sewers are typically much smaller than combined sewers, and they are not designed to transport storm water.
  - Backups of raw sewage can occur if excessive infiltration/inflow (dilution by storm water and/or groundwater) is allowed into a sanitary sewer system.

- Communities that have urbanized in the mid-20th century or later generally have built separate systems for sewage (sanitary sewers) and storm water, because precipitation causes widely varying flows, reducing sewage treatment plant efficiency.

- As rainfall travels over roofs and the ground, it may pick up various contaminants including:

  - Soil particles and other sediment
  - Heavy metals
  - Organic compounds
  - Animal waste
  - Oil
  - Grease

- Some jurisdictions require storm water to receive some level of treatment before being discharged directly into waterways.

  - Examples of treatment processes used for storm water include retention basins, wetlands, buried vaults with various kinds of media filters, and vortex separators (to remove coarse solids).

Additional Resources

Internet sites represent a vast resource of information. Using one of the search engines on the Internet such as Google or DuckDuckGo, find more information by searching for words, phrases, or websites: Always use caution when searching for information on the Internet. Follow guidelines to ensure the accuracy and reliability of the information you find.

Search words: particulate organic matter, sewage treatment systems, colloidal particles, dissolved inorganic material, storm water runoff,

Websites: www.usgs.gov, www.epa.gov

Videos: https://www.youtube.com/watch?v=CoFuQZBPCKo , https://www.youtube.com/watch?v=FvPakzqM3h8

Self-Assessment

1. Describe the steps in sewage treatment.

2. Modern day health is attributed to sewage collection systems and effective sewage treatment, among other things. What pandemic was thought to have contributed to spreading by raw sewage?

3. List the types of sewage that are generated by households, industry, and commerce.

4. Describe how storm water runoff affects the sewer system.

19 Removing Pollutants from Sewage

Major Concept

Sewage treatment uses physical, chemical, and biological processes to remove physical, chemical, and biological contaminants and to produce an environmentally safe fluid waste stream (water) and a solid waste.

Objectives

- Explain the process of removing pollutants from sewage
- Describe an activated sludge system
- List the types of water disinfecting techniques

Key Terms

- Activated sludge
- Activated sludge system
- Bar screen
- Biological Nutrient Removal (BNR)
- Biological Oxygen Demand (BOD)
- Biological treatment
- Chlorine gas
- Floc
- Grit chamber
- Primary clarifier
- Raw sludge

Chapter Resource

Complementary *full color* illustrations, photos, charts, and graphs are available for Chapter 19 by following this URL: https://tinyurl.com/7vxzdaks This digital resource will enhance your understanding of the chapter concepts.

Preliminary Treatment (Removal of Debris and Grit)

- Because debris and grit will damage or clog pumps and later treatment processes, removing them is a necessary step.

- It involves two steps, a screening out of debris and a settling of grit.

  o Debris is removed by letting raw sewage flow through a **bar screen**, a row of bars mounted about 1 inch (2.5 cm) apart.

    ✓ Debris is mechanically raked from the screen and taken to an incinerator.

  o The wastewater then passes through a grit-settling tank (**grit chamber**), a swimming pool-like tank where its velocity is slowed just enough to permit the grit to settle.

    ✓ The settled grit is mechanically removed from these tanks and taken to landfills.

Primary Treatment

- Primary treatment is the removal of particulate organic material.

- After preliminary treatment, the water moves on to primary treatment where it flows very slowly (2 meters per hour) through large tanks called **primary clarifiers** because it flows slowly through these tanks, the water is nearly motionless for several hours.

    - The particulate organic material (about 30 – 50% of the organic material) settles to the bottom, where it can be removed.
    - At the same time, fatty material floats to the top where it is skimmed from the surface.
    - All the material removed, both particulate organic and fatty materials is combined into what is referred to as **raw sludge**.

Secondary Treatment

- Secondary treatment is the removal of colloidal and dissolved organic material.

- This is also called **biological treatment** because it makes use of organisms.

- Basically, an environment is created to enable these organisms to feed on the colloidal and dissolved organic material and break them down to carbon dioxide and water via their cellular respiration.

    - The sewage water from primary treatment is the food and water rich medium. The only thing that needs to be added is oxygen to enhance the organisms' respiration and growth.
    - The amount of dissolved oxygen needed by these aerobic decomposers to break down organic materials in a given volume of water at a certain temperature over a specific period is referred to as the **biological oxygen demand (BOD)**.

BOD Level in mg/liter	Water Quality
1 - 2	**Very Good:** There will not be much organic matter present in the water supply.
3 - 5	**Fair:** Moderately Clean
6 - 9	**Poor:** Somewhat Polluted - Usually indicates that organic matter present and microorganisms are decomposing that waste.
100 or more	**Very Poor:** Very Polluted - Contains organic matter.

- The **activated-sludge system** may be used. In this system, water from primary treatment enters a long tank that is equipped with an air bubbling system.

- A mixture of detritus-feeding organisms, referred to as **activated-sludge**, is added to the water as it enters the tank, and the water is vigorously aerated as it moves through the tank.

    - Organisms in this well-aerated environment reduce the biomass of organic material, including pathogens, as they feed.

        - ✓ As organisms feed on each other, they tend to form clumps, referred to as **floc**, that settles readily when the water is motionless.

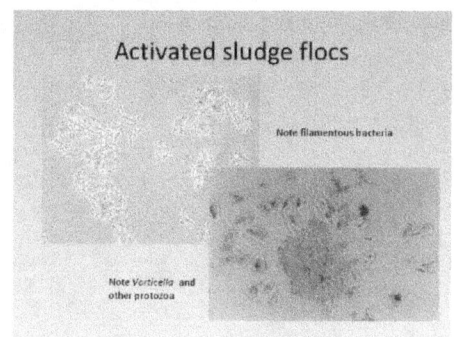

Activated sludge flocs

Note filamentous bacteria

Note Vorticella and other protozoa

- ✓ Thus, from the aeration tank, the water is passed into a secondary clarifier tank where the organisms settle out.

  - o The water is now cleaner as 90% of the organic material removed flows on.
  - o The settled organisms are pumped back into the entrance of the aeration tank.

    - ✓ They are the activated sludge that is added at the beginning of the process.
    - ✓ Surplus amounts of activated sludge, which occur as populations of organisms grown, are removed and added to raw sludge.

Biological Nutrient Removal

- Biological nutrient removal involves the removal of dissolved inorganic material.

- Today, with increased knowledge of problems of cultural eutrophication, secondary activated-sludge systems are being modified and operated in a manner that achieves nutrient removal as well as detritus oxidation, a process known as **biological nutrient removal (BNR).**

- In natural nitrogen cycle, nutrient forms of nitrogen are converted by various bacteria back to nonnutritive nitrogen gas in the atmosphere. This process is called denitrification.

  - o For biological removal of nitrogen, the activated sludge system is partitioned into zones, and the environment in each zone is controlled in a manner to promote the denitrifying process.

- With respect to phosphate, in an environment that is rich in oxygen but relatively lacking in food, the environment of zone 3, bacteria take up phosphate from the solution and store it in their bodies.

- Thus, phosphate is removed as the excess organisms are removed from the system.

  - o These organisms with the phosphate they contain, are added to, and treated with the raw sludge, ultimately producing a more nutrient rich treated sludge (biosolids – organic materials removed from sewage effluents in the cause of treatment) product.

- As an alternative to BNR, various chemical processes may be used.

  - o One is simply to pass the effluent from standard secondary treatment through a filter of lime, which causes the phosphate to precipitate out as insoluble calcium phosphate.

Final Cleansing and Disinfection

- After biological nutrient removal, the wastewater is subjected to a final cleansing by filtration through a bed of sand and by disinfection.

- Although few pathogens survive the combined stages of treatment and sand filtration, public-health rigors still demand that the water be disinfected before discharge into natural waterways.

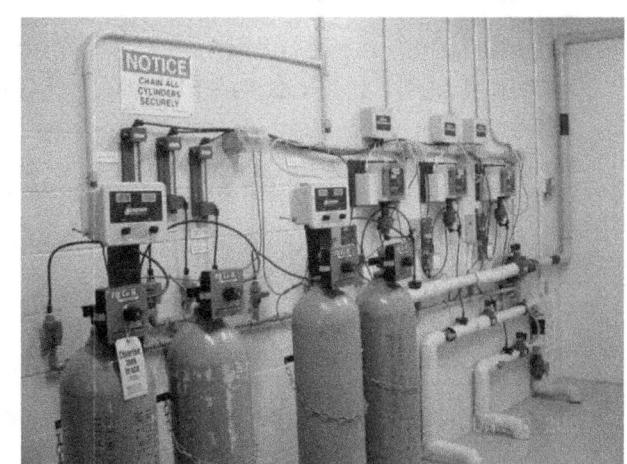

- The most widely used disinfection agent is **chlorine gas** because it is both effective and relatively inexpensive.

  - However, this treatment also introduces chlorine into natural waterways, and even minute levels of chlorine can harm aquatic animals.

    ✓ For instance, the hatching of trout eggs and development of the embryos is affected by the presence of chlorine.
    ✓ In addition, chlorine reacts spontaneously with organic compounds to some extent to form chlorinated hydrocarbons, organic molecules with chlorine atoms attached.

  - Many of these compounds are toxic and non-biodegradable and some have been identified as compounds that cause cancer, abnormal development, and reproductive problems.

  - Because of the negative side effects of chlorine, an additional agent may be added to convert chlorine to a chemically inactive form.

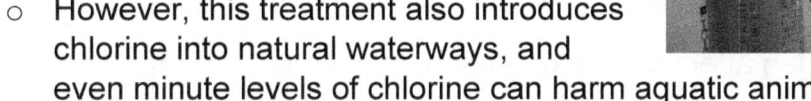

- Other disinfecting techniques that are coming into use as an alternative disinfecting agent is the ozone gas, which is extremely effective in killing microorganisms and in the process break down to oxygen gas, which actually improves water quality. However, because ozone is unstable and hence explosive, it must be generated at the point of use, a step that demands considerable capital investment and energy.

- Another disinfecting technique is to pass the effluent through an array of ultraviolet lights mounted in the water (standard fluorescent lights without the white coating emit ultraviolet).

  - The ultraviolet radiation kills microorganisms but does not otherwise affect water.

- After these steps of treatment, the wastewater has a lower organic and nutrient content than most bodies of water into which it is discharged.

Additional Resources

Internet sites represent a vast resource of information. Using one of the search engines on the Internet such as Google or DuckDuckGo, find more information by searching for words, phrases, or websites: Always use caution when searching for information on the Internet. Follow guidelines to ensure the accuracy and reliability of the information you find.

Search words: Grit-settling tank, bar screen, raw sludge, biological treatment, chlorine gas, ozone gas

Self-Assessment

1. Raw sludge is a product of which level of sewage treatment?

2. How does the level of oxygen affect the decomposers in secondary treatment?

3. What is considered the safest method of disinfecting wastewater before being discharged into a body of water?

4. Describe biological nutrient removal (BNR).

20 Major Air Pollutants and Their Impacts

Major Concept

Air pollution consists of chemicals or particles in the air that can harm the health of humans, animals, and plants and it can damage structures. Pollutants can be gases, solid particles, or liquid droplets.

Objectives

- List the nine major air pollutants
- Explain the impact that each pollutant has on the environment

Key Terms

- Ozone shield

Chapter Resource

Complementary *full color* illustrations, photos, charts, and graphs are available for Chapter 20 by following this URL: https://tinyurl.com/7vxzdaks This digital resource will enhance your understanding of the chapter concepts.

Air Pollutants

- The following nine pollutants have been identified as the most widespread and serious.

    - Suspended Particulate Matter

        ✓ This is a complex mixture of solid particles and aerosols (liquid particles) suspended in the air.
        ✓ They are seen as dust, smoke, and haze.
        ✓ The particles impair many respiratory functions, especially in individuals with chronic respiratory problems.
        ✓ Particles less than 10 micrometers have the greatest effect on health because it can be inhaled.

        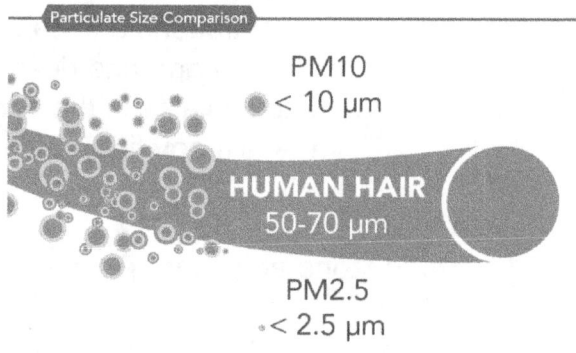

    - Volatile Organic Compounds (VOCs)

        ✓ These include materials such as gasoline, paint solvents, and organic cleaning solutions, which evaporate and enter the air in a vapor state as well as fragments of molecules resulting in the incomplete oxidation of fuels and wastes.

- ✓ VOCs are major factors in the formation of photochemical smog.

- o Carbon Monoxide (CO)

  - ✓ This is an invisible odorless gas that is highly poisonous to air-breathing animals because of its ability to block the delivery of oxygen to the organs and tissues.

- o Nitrogen Oxides (NOx)

  - ✓ These include several nitrogen-oxygen compounds (all gases).
  - ✓ They are converted to nitric acid in the atmosphere and are a major source of acid deposition.
  - ✓ Nitrogen dioxide is a lung irritant that can lead to acute respiratory disease in children.

- o Sulfur Oxides (SOx)

  - ✓ Sulfur dioxide (SO_2) is a gas that is poisonous to both plants and animals.
  - ✓ Children and the elderly are especially sensitive to SO_2.
  - ✓ It is converted into sulfuric acid in the atmosphere and is a major source of acid deposition.

- o Lead and Other Heavy Metals

  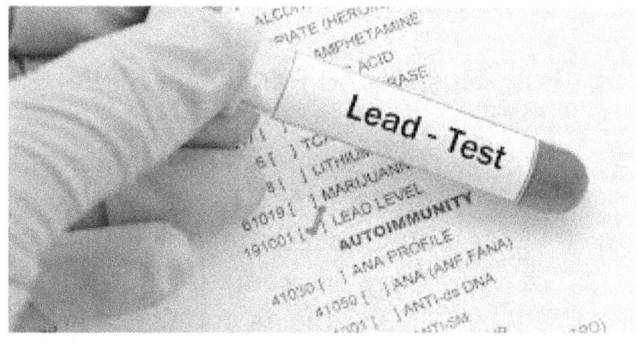

  - ✓ Lead and other heavy metals are very dangerous at low concentrations and can lead to brain damage and death.
  - ✓ It accumulates in the body and impairs many tissues and organs.

- o Ozone and Other Photochemical Oxidants

  - ✓ **Ozone shield** (the layer of ozone gas – O_3) in the upper atmosphere is needed to screen out harmful ultraviolet radiation from the sun.
  - ✓ It may also be used for disinfecting water.
  - ✓ Ozone is a pollutant in the lower atmosphere.
  - ✓ It is highly toxic to both plants and animals.
  - ✓ In animals, it damages lung tissues and implicated in many lung disorders.

- Air Toxics and Radon

  - ✓ Toxic chemicals in the air include carcinogenic chemicals, radioactive materials, and other chemicals (such as Asbestos, Vinyl chloride, Benzene) that are emitted as pollutants.
  - ✓ Radon is a radioactive gas produced by natural processes within Earth.
  - ✓ All radioactive substances have the potential to damage any living matter they come in contact with.

- Odor

  - ✓ Unpleasant odors from industry wastes and livestock wastes from concentrated animal feeding operations (CAFOs) can, in a way, be considered pollution.
  - ✓ These airborne emissions, including ammonia, hydrogen sulfide (H_2S) and volatile organic compounds originate from settling ponds, holding ponds, confinement buildings, waste storage areas, and land application of animal waste.

A type of Confined Animal Feeding Operations (CAFOs)

Additional Resources

Internet sites represent a vast resource of information. Using one of the search engines on the Internet such as Google or DuckDuckGo, find more information by searching for words, phrases, or websites: Always use caution when searching for information on the Internet. Follow guidelines to ensure the accuracy and reliability of the information you find.

Self-Assessment

Visit the website. https://airquality.deq.idaho.gov/home/map Click through the Interactive Maps to see Real-Time Air Monitoring for the state of Idaho. Choose an area of interest and share data with the class.

21 Soil and Soil Ecosystem

Major Concept

Soil is a dynamic ecosystem system involving three components: mineral particles, detritus, and organisms feeding on detritus.

Objectives

- Describe the purpose of soil to plants
- List the attributes of soil
- Indicate how a soil triangle is used to determine the type of soil
- Explain nutrient and water-holding capacity

Key Terms

- Compacted
- Fertilizer
- Humus
- Inorganic fertilizer
- Irrigation
- Leaching
- Loam
- Organic fertilizer
- Stomata
- Topsoil
- Transpiration
- Water holding capacity
- Weathering

Chapter Resource

Complementary *full color* illustrations, photos, charts, and graphs are available for Chapter 21 by following this URL: https://tinyurl.com/7vxzdaks This digital resource will enhance your understanding of the chapter concepts.

Soil and Plants

- Soil is a dynamic system involving three components: mineral particles, detritus, and organisms feeding on detritus.

- Plants need an environment that supplies optimal amounts of the mineral nutrients, water, and air (oxygen).

- The pH (relative acidity) and salinity (salt concentration) of the soil are also important.

- The soil's ability to support plant growth is called soil fertility (the presence of proper amounts of nutrients).

Mineral Nutrients and Nutrient-Holding Capacity

- Mineral nutrients – phosphate(PO_4^{3-}), potassium (K^+), calcium (Ca^{2+}), and other ions – are present in various rocks along with non-nutrient elements.

- Through the process of **weathering**, they initially become available to the plant roots.

  o Weathering is the gradual breakdown of rock into smaller and smaller particles, caused by natural, chemical, physical, and biological factors.

- Mostly, the nutrients that support plant growth in natural ecosystems are supplied through the breakdown and release (recycling) from detritus.

- Nutrients may be lost through **leaching** (the process of removing materials from the soil by water).

  o Leaching not only lessens soil fertility, it also contributes to pollution when materials removed from the soil enter waterways.

  o For this reason, and is called nutrient-holding capacity.

- Agricultural systems do require inputs of nutrients to replace those removed with the harvest.

- Nutrients are replenished with applications of **fertilizer**, material that contains one or more of the necessary nutrients.

- Fertilizer may be organic or inorganic.

  o **Organic fertilizer** includes plant or animal wastes or both. Manure and compost (rotted organic material) are two examples.

  o **Inorganic fertilizers** are chemical formulations or required nutrients without any organic matter included.

    ✓ Inorganic fertilizers are much more prone to leaching than are organic fertilizers.

Water and Water-Holding Capacity

- Plants need water. A constant stream of water, absorbed by the roots, passes through the plant, and exits as water vapor through microscopic pores in the leaves – a process called **transpiration.**

  o The pores called **stomata** are also essential to permit the entry of carbon dioxide and the exit of oxygen in photosynthesis.

- Inadequate water results first in wilting, conditions that conserve water, and also shuts off photosynthesis by closing the stomata and preventing gas exchange.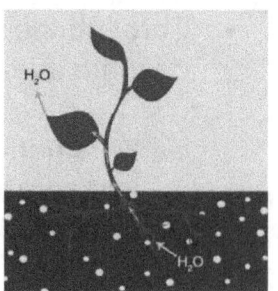
    - If the wilted condition is too severe or too prolonged, the plants die.
    - In this situation, water may be supplied artificially by **irrigation**.

Attributes of Soil

- Infiltration

    - This is the passing of water through soil. If water runs off the ground surface it will not be useful to plants and worse, it may cause erosion.

- **Water holding capacity**

    - This is the ability of a soil to hold water so that it will be available to plants.

    - Poor water holding capacity implies that most of the infiltration water percolates deeper below the reach of the roots, which becomes useless to plants since they cannot use it.

    - What is desired is a good water holding capacity, the ability to hold a large amount of water like a sponge, providing a reservoir on which plants can draw between rains.

- Evaporative water loss

    - High rate of water evaporation from the soil surface will deplete the soil's water reservoir without serving the needs of plants.

- Soil aeration

    - Land plants depend on the soil being loose and porous enough to allow the diffusion of oxygen into and carbon dioxide out of the soil.

    - Oxygen is required by plants for metabolic processes and this oxygen is obtained from the soil air (for most land plants).

    - When soil is **compacted** it lacks the oxygen needed by plants, reduces infiltration, and increases runoff.

- Relative acidity (pH)

    - The term pH refers to the acidity or alkalinity (basicity) of any solution.

    - A solution that is neither acidic nor alkaline is said to be neutral and has a pH of 7.

- The pH scale runs from 0 to 14.

- Most plants (as well as animals) require a pH near neutral, and most natural environments provide this.

Salt and Water Uptake

- High salt concentration in the soil makes it impossible for roots to take in water.

- If the salt concentration in the soil is very high, water can be drawn out of the plant, resulting in dehydration and death.

  - In summary, to support a good crop the soil must:

    ✓ Have a good supply of nutrients and nutrient-holding capacity
    ✓ Allow infiltration and have good water-holding capacity and resist evaporative water loss
    ✓ Have a porous structure that permits good aeration
    ✓ Have a pH near neutral
    ✓ Have a low salt content

Soil Texture

- This refers to the relative size of the mineral particles that make up the soil.

- Soil texture is generally in terms of the sand, silt, and clay content.

  - If any of the sand, silt, or clay makes a higher proportion of the soil, we speak of a sandy, silty, or clayey soil.

  - A proportion that is commonly found consists of roughly 40% sand, 40% silt, and 20% clay.

    ✓ A soil with such proportion is called a **loam**.

  - A soil triangle is used to determine the type of soil.

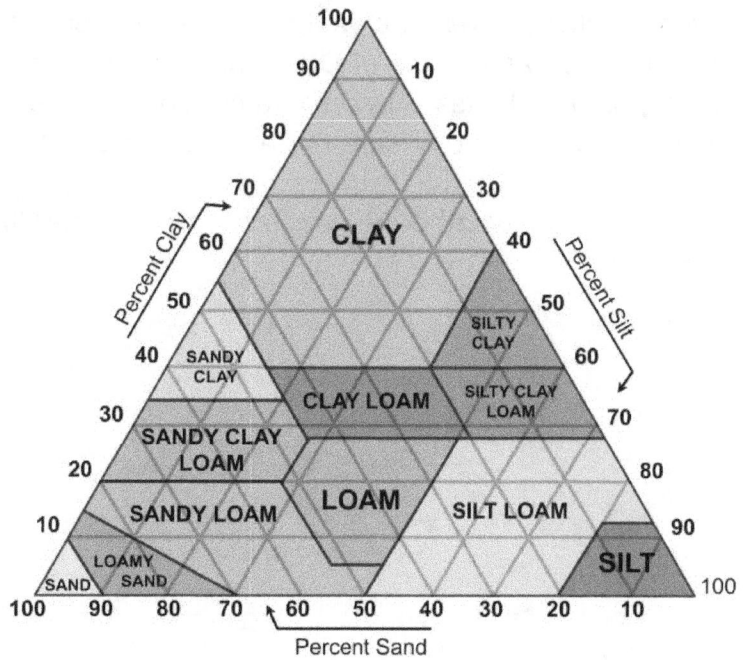

Detritus, Soil Organisms, Humus, and Topsoil

- The accumulation of dead leaves, roots, and other detritus on and in the soil supports a complex food web, including numerous species of:

  - Bacteria
  - Fungi
  - Protozoan
  - Mites
  - Insects
  - Millipedes
  - Spiders
  - Centipedes
  - Earthworms
  - Snail
  - Slugs
  - Moles
  - Other burrowing animals

- As these organisms feed, the bulk of the detritus is consumed through their cell respiration, and carbon dioxide, water, and mineral nutrients are released as byproducts.

  - However, each organism leaves a certain portion undigested – that is, a certain portion resists breakdown by the organism's digestive enzyme.

  - This residue is partly decomposed organic matter is called **humus**, which makes a part of the topsoil.

- **Topsoil** is the surface layer of soil, which is rich in humus and other organic material, both living and dead.

  - As a result of the activity of organisms living in the topsoil, it generally has a loose, crumbly structure as opposed to being a compact mass.

Additional Resources

Internet sites represent a vast resource of information. Using one of the search engines on the Internet such as Google or DuckDuckGo, find more information by searching for words, phrases, or websites: Always use caution when searching for information on the Internet. Follow guidelines to ensure the accuracy and reliability of the information you find.

Search words: weathering, leaching, types of fertilizers, transpiration, saline and sodic soils, soil texture

Self-Assessment

Use the soil texture triangle to name soils. Determine the appropriate soil textural class using the soil textural triangle

Particle Type and Percentage	Appropriate Soil Textural Class
40% Sand 50% Silt 10% Clay	
70% Sand 15% Silt 15% Clay	
35% Sand 15% Silt 50% Clay	
20% Sand 60% Silt 20% Clay	
30% Sand 40% Silt 30% Clay	

Now complete this chart using the soil textural triangle.

% Sand	% Silt	% Clay	Texture
5		50	
27	35		
	31	33	
22	23		
10		7	
	7	23	

22 Pests and Pest Control

Major Concept

Pests are organisms that are noxious, destructive, or troublesome to humans, directly or indirectly and they are controlled by various methods.

Objectives

- List the types of pests and the negative role they play for humans
- Describe how pests can be chemically controlled and the potential problems associated with chemical use
- Identify alternatives to chemical pesticides
- Outline the steps in Integrated Pest Management (IPM)

Key Terms

- Integrated Pest Management (IPM)
- Pesticides
- Pests

Chapter Resource

Complementary *full color* illustrations, photos, charts, and graphs are available for Chapter 22 by following this URL: https://tinyurl.com/7vxzdaks This digital resource will enhance your understanding of the chapter concepts.

Definition and Importance of Pests

- **Pests** are defined as organisms that are noxious, destructive, or troublesome to humans, directly or indirectly.

    o A better way to understand this definition is by expanding the negative role described above into the following categories:

    ✓ Organisms that cause disease in humans or domestic plants and animals. These pests include viruses, bacteria, and parasitic organisms like intestinal organisms and flukes.
    ✓ Organisms that annoy people and domestic animals and that may transfer disease by biting or stinging. For example: flies, ticks, bees, and mosquitoes.
    ✓ Organisms that feed on ornamental plants or agricultural crops, e.g., snails, slugs, rats, mice, certain birds, and worms.
    ✓ Animals that attack and kill domestic animals e.g., coyotes, foxes, and racoons.
    ✓ Organisms that cause wood, leather, and other materials to rot and food to spoil e.g., bacteria and fungi.

- ✓ Plants that compete with agricultural crops, forest, and forage grasses for light and nutrients.

  o All these categories of pests are important in human affairs.

  o We try to bring these pests under control mainly for the purpose of protecting our food, health, and other aspects of our well-being.

- Pests are controlled by using **pesticides** which include chemicals that kill animals, insects, worms, and rodents considered pests (insecticides, nematocides, rodenticides) and chemicals that kill weeds (herbicides) and fungus (fungicide).

- Problems associated with the use of chemical pesticides:

  o Development of resistant pests
  o Resurgence and secondary pest outbreaks
  o Adverse environmental and human health effects

Alternative to Chemical Pesticide

- Scientific findings suggest ways in which pest population may be controlled without resorting to synthetic chemical pesticides.

- They are:

  o Cultural control
  o Control by natural enemies
  o Genetic control
  o Natural chemical control

Integrated Pest Management (IPM)

- **Integrated pest management (IPM)** uses a combination of methods to control pests that result in a minimum of damage to the environment.

  o The control selected will:

    ✓ Cause the least amount of injury to the pest's natural enemies
    ✓ Have the least potential to cause harm to humans
    ✓ Cause the least damage to the environment
    ✓ Be the most cost-effective
    ✓ Be the easiest to carry out effectively

- IPM involves using a wide range of pest control strategies, making decisions requires good information.

- The key pests must be identified. These are the pests that regularly causes losses in crops.

- Biological characteristic of the crop and of the key pest are important in using IPM.
- Practices can be selected to promote crop growth and disrupt reproduction and growth of the pest.

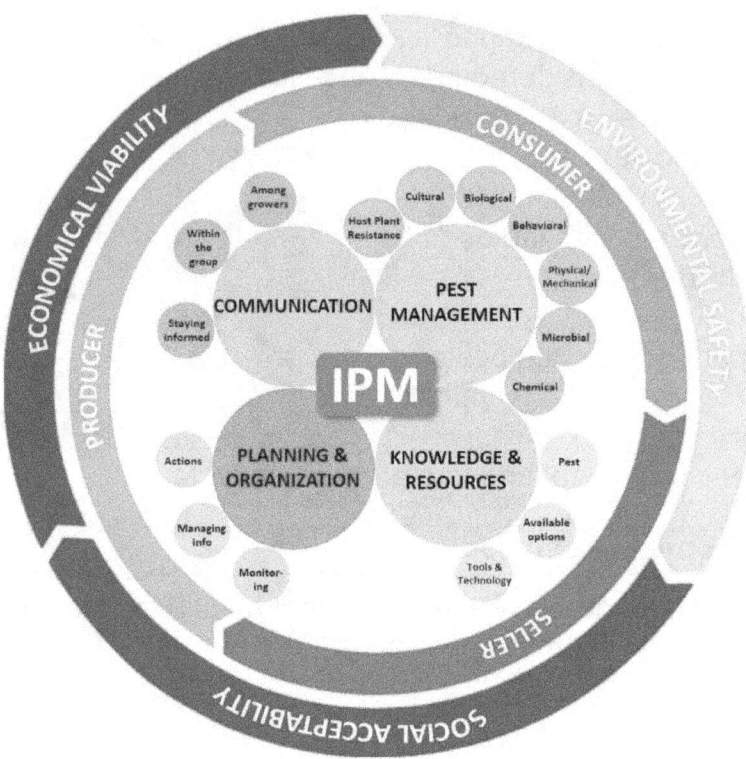

Additional Resources

Internet sites represent a vast resource of information. Using one of the search engines on the Internet such as Google or DuckDuckGo, find more information by searching for words, phrases, or websites: Always use caution when searching for information on the Internet. Follow guidelines to ensure the accuracy and reliability of the information you find.

Search words: insecticides, pesticides, nematocides, rodenticides, herbicides, fungicides

Self-Assessment

1. Give an example of the organisms that are considered pests and why.

2. Describe the different types of pesticides and what they are used to control.

3. Name three alternatives to chemical pesticides.

23 Global Human Population

Major Concept

At the present rate of growth, the world population will be approximately 9+ billion by the year 2050.

Objectives

- List the reasons why there was a slow population growth in the past
- Explain the reasons for the population explosion
- Identify the current trend in population growth
- Indicate how the current population growth is affecting the environment

Key Terms

- Thomas Robert Malthus
- Negative Population Growth (NPG)

Chapter Resource

Complementary *full color* illustrations, photos, charts, and graphs are available for Chapter 23 by following this URL: https://tinyurl.com/7vxzdaks This digital resource will enhance your understanding of the chapter concepts.

The Trend

- From the beginning of human history to the beginning of the 1800s, the human population increased slowly and variably with periodic setbacks.

- It was in the 1830s that the human population reached 1 billion.

- Since then, the human population has been increasing steadily and is currently estimated to be 7 billion.
- Its rate of growth is also estimated to be 88 million people per year.

- It is projected, mathematically, that at the present rate of growth, the population will be approximately 9+ billion by the year 2050.

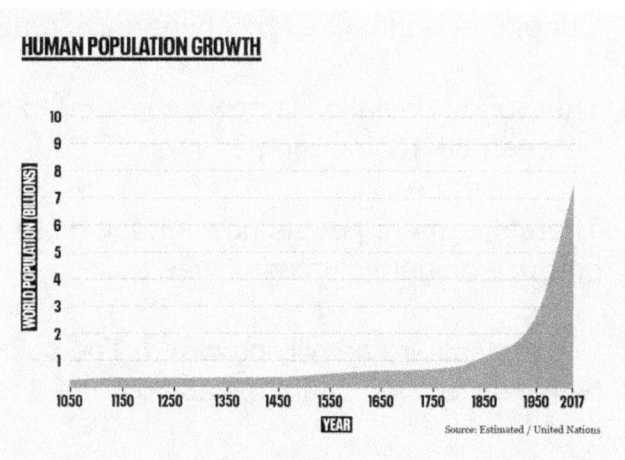

Reasons for Population Changes

- The main reason for slow and fluctuating population growth prior to early 1800s were the prevalence of diseases such as smallpox, diphtheria, measles, and scarlet fever, which were often fatal.

    o These diseases mostly affect infants and children.
    o Besides, epidemics of diseases such as typhoid fever and cholera, and the black plague of the 14th century, eliminated large numbers of adults.
    o Famine caused by environmental factors such as droughts were also a limiting factor in population growth.

- Prior to the 1800s, biologically speaking, the population was essentially in a dynamic balance with natural enemies – mainly due to diseases, and other aspects of environmental resistance.

    o High reproductive rates were balanced by high mortality, especially among infants and children.
    o With high birth and death rates, the population growth rate was low in those societies.

- In the 1800s, Louis Pasteur and other scientists made the major discovery that infectious agents such as bacteria, viruses, and parasites caused diseases.

- This was followed by techniques of providing protection by means of vaccinations and use of antibiotics.

    o With those discoveries, came major improvements in sanitation and personal hygiene.
    o Improvement in nutrition was also advancing.
    o Therefore, better medicine, sanitation, and nutrition reduced death rates among infants and children, while birth rates remained high.
    o Again, in biological sense, the human population entered into exponential growth, as a result of the natural population being freed from natural enemies and other environmental restraints.

Population and the Environment

- Our planet is inhabited by an increasing number of people as years go by.

- The world population is now estimated to be more than 7 billion and is currently growing between 88-100 million per year.

- Therefore, more people now inhabit more extensive areas, consumes more products, and produce bigger volumes of wastes.

- Some cities are heavily crowded. The solutions to population problems are different for highly developed countries as compared to less developed countries.

- The population problem is especially serious in developing countries because they have about 75% of the world population and only 20% of the wealth.
- The use of extensive areas for human purpose has led to natural habitat alterations and hence biodiversity loss.
- Advancing technology has created industries that produce goods for human needs, and at the same time they generate by-products that pollute the terrestrial, aquatic, and ambient environments.
- We will study the population problems in detail later.

Promoting the Development of Low-income Nations

- Increasingly world leaders realize and publicly state that population growth, poverty, and environmental degradation is interrelated.

- Borrowing from the World Bank and other rich countries for long-term and sustainable development programs can help.

  - Programs/projects funded by other international/national organizations including charitable organizations can play an important role to increase or improve low-income nations.

- World Bank funded programs have increased the gross national products of some countries.

- Together with organizations such as the United Nations (UN), the World Health Organization (WHO), Food and Agricultural Organization (FAO), UN Educational, Scientific, and Cultural Organization (UNESCO), and United Children Fund (UNICEF), including other charitable organizations, some improvements in social issues, literacy, water, health, education, and sanitation have been made.

  - However, the progress made is little.
  - More than 1.3 billion people still remain in absolute poverty, illiterate, and without access to clean water or adequate nutrition.
  - Environmental degradation is rampant, and fertility rates remain unacceptably high.

A New Direction for Development

- To break the cycle of poverty, high fertility, and environmental degradation in the poor nations, effort should be directed on the following:
  - Education – especially improving literacy and educating girls and women equally with boys and men.
  - Improving health – especially lowering infant mortality.
  - Making contraceptives available
  - Enhancing income

        - Improving resource management (reversing environmental degradation)
- Education and training of women should have special focus because they are children bearers and providers of nutrition, childcare, hygiene, and early education.

Additional Resources

Internet sites represent a vast resource of information. Using one of the search engines on the Internet such as Google or DuckDuckGo, find more information by searching for words, phrases, or websites: Always use caution when searching for information on the Internet. Follow guidelines to ensure the accuracy and reliability of the information you find.

Websites: https://en.wikipedia.org/wiki/List_of_natural_disasters_by_death_toll
https://www.worldometers.info/world-population/

Assessment

1. Reading: Negative Population Growth (NPG; Text of an email received 3 February 2015)

Based on your reporting on college and university education, I thought you might be interested in a new Press Release issued today by NPG (http://www.npg.org/)

NPG is once again offering our Annual Scholarship Contests to high school seniors and currently enrolled college students. Winners will receive Scholarships ranging from $500 to $2,500 paid towards their undergraduate tuition. Each year, NPG makes this substantial investment, helping America's next generation of leaders to meet the rising costs of education.

NPG is a national nonprofit membership organization dedicated to educating the American public and political leaders regarding the damaging effects of population growth. We believe that our nation is already vastly overpopulated in terms of the long-range carrying capacity of its resources and environment. The evidence is all around us: millions are out of work, schools and hospitals are overcrowded, our natural resources are dwindling, pollution is increasing, species are threatened or have vanished forever, and our infrastructure and social services are crumbling under the strain. We simply cannot afford to continue our growth. It is unsustainable, and irresponsible policies are the driving force behind it.

I hope you will take a moment to review our release, and please feel free to contact me with any questions you may have.

Best,
Tracy Canada
Deputy Director, NPG
Working together for a livable future.

- About NPG

Negative Population Growth, Inc. (NPG) is a national nonprofit membership organization. It was founded in 1972 to educate the American public and political leaders about the devastating effects of overpopulation on our environment, resources and standard of living. We believe that our nation is already vastly overpopulated in terms of the long-range carrying capacity of its resources and environment.

We urgently need a National Population Policy with the goal of eventually stabilizing our population at a sustainable level, far below today's.

Most politicians, big business and its supporting economists call for growth as a solution to all our problems. Apparently, they believe in perpetual growth, which is a mathematical absurdity on a finite planet. There must be limits. Science is demonstrating that human population and consumption in the United States and the world are already too large and are destroying the natural systems that support us. We must not simply stop population growth; we must turn it around.

Since 1972, NPG has been making that case. We do not simply identify the problems, we propose solutions.

- For the Record: This Is Not a New Idea

The Reverend Thomas Robert Malthus FRS (13 February 1766 – 29 December 1834) was an English cleric and scholar, influential in the fields of political economy and demography.

His "An Essay on the Principle of Population" observed that sooner or later population will be checked by famine and disease, leading to what is known as a Malthusian catastrophe. He wrote in opposition to the popular view in 18th-century Europe that saw society as improving and in principle as perfectible. He thought that the dangers of population growth precluded progress towards a utopian society: "The power of population is indefinitely greater than the power in the earth to produce subsistence for man". As a cleric, Malthus saw this situation as divinely imposed to teach virtuous behavior. Malthus wrote:

  o That the increase of population is necessarily limited by the means of subsistence,
  o That population does invariably increase when the means of subsistence increase, and,
  o That the superior power of population is repressed, and the actual population kept equal to the means of subsistence, by misery and vice.

Malthus placed the longer-term stability of the economy above short-term expediency. He criticized the Poor Laws, and (alone among important contemporary economists) supported the Corn Laws, which introduced a system of taxes on British imports of wheat. His views became influential, and controversial, across economic, political, social and scientific thought.

And they continue to resurface and be taught. How do you feel about this?

2. Population Scenarios – a Class Discussion
- Some claim that population growth is beneficial because people provide ideas, creativity and work. The greatest technological advances and improvements in living standards that have occurred in parallel with the population explosion of the past 200 years support this belief.

- Some take the position that any artificial interference in the reproductive process (including the use of sex education, contraceptive, and especially abortions) is immoral. Therefore, for them, any thought of altering the course of population growth, except by abstinence, is not debatable.
- Some argue that population growth is not the issue as much as consumption is, at least for the present. What needs to be achieved more than reducing in numbers, they say is adopting conservation measures that reduce consumption.
- Others take the position that population growth will level off by itself well within the capacity to support the population. This view is supported by the facts that we have been able to expand agricultural production even faster than population growth and that the average number of births per woman is coming down.
- Still other are disturbed by the fact that population programs often seem to have the trappings of social engineering - the rich trying to get rid of the poor or minorities by preventing them from having children.

These scenarios are complex to discuss exhaustively at your level. Whatever stand an individual takes, one thing is certain, he/she will be responsible for the actions he/she takes.

24 Pollution from Hazardous Chemicals

Major Concept

Hazardous material is anything that can cause injury, disease, or death to humans; damage to property; degradation of the environment. These materials possess one or more of the following characteristics: ignitability, corrosivity, reactivity, and toxicity.

Objectives

- List the four characteristics of hazardous materials
- Explain why heavy metals are toxic to the environment
- Describe the hazards of nonbiodegradable synthetic organics
- Compare bioaccumulation and biomagnification

Key Terms

- Bioaccumulation
- Biomagnification
- Carcinogenic
- Chlorinated hydrocarbons (organic chlorides)
- Corrosivity
- Halogenated hydrocarbons
- Ignitability
- Mutagenic
- Nonbiodegradable
- Reactivity
- Teratogenic
- Toxicity

Chapter Resource

Complementary *full color* illustrations, photos, charts, and graphs are available for Chapter 24 by following this URL: https://tinyurl.com/7vxzdaks This digital resource will enhance your understanding of the chapter concepts.

Hazardous Material (HAZMAT)

- Hazardous material (HAZMAT) is anything that can cause injury, disease, or death to humans; damage to property; degradation of the environment.

- The Environmental Protection Agency (EPA) categorizes substances on the basis of hazardous properties:

  - **Ignitability**

    ✓ Substances that catch fire readily e.g., gasoline, alcohol.

  - **Corrosivity**

    ✓ Substances that corrode storage and equipment e.g., acids, sodium hydroxide.

- **Reactivity**
  - ✓ Substances that are chemically unstable and may explode or create toxic fumes when mixed with water e.g., phosphorus, sulfuric acid.
- **Toxicity**
  - ✓ Substances that are injurious to health when ingested or inhaled e.g., chlorine, ammonia, pesticides, and formaldehyde.
- In brief, hazardous materials are any materials having one or more of the following characteristics: ignitability, corrosivity, reactivity, and toxicity.

Toxic Chemicals

- Toxic chemicals that present a long-term threat to human or the environment may be classified as:

- Heavy Metals

  - These are any of the high atomic weight metals such as lead, mercury, cadmium, zinc, tin, chromium, and copper.

  - All may be serious pollutants in water or soil because they are toxic in relatively low concentrations and they tend to **bioaccumulate** (accumulation of higher and higher concentration of potentially toxic chemicals in organisms).

  - These metals are widely used in industries, especially in metal or metal-plating shops and in such products as batteries and electronics.

  - They are also used in certain pesticides and medicines.

  - Because they are heavy metal compounds, they can have brilliant colors and are therefore used in paint pigments, glazes, ink, and dye.

  - Thus, heavy metals may enter the environment whenever any of these products are produces, used, and discarded.

  - Heavy metals are extremely toxic because, as ions or in certain compounds, they are soluble in water and may be readily absorbed into the body, where they tend to combine with and inhibit the functioning of particular vital enzymes.

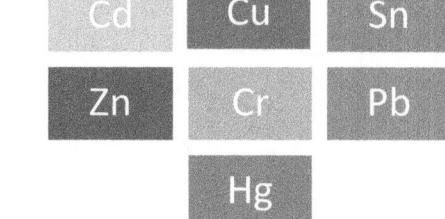

  - Even very small amounts can have severe physiological or neurological consequences.

  - The mental retardation caused by lead poisoning and the insanity and crippling birth defects caused by mercury are good examples.

- Nonbiodegradable Synthetic Organics

- **Nonbiodegradable** is a material that cannot be consumed or broken down by biological organisms.

  o They include plastics, aluminum, and many chemicals used in industrial and agricultural sectors.

  o Nonbiodegradable synthetic organics include:

    ✓ Fibers
    ✓ Synthetic rubber
    ✓ Modern paints/coatings
    ✓ Solvents
    ✓ Pesticides
    ✓ Wood preservatives

  o The advantage of nonbiodegradable materials to humans is that fungi or bacteria cannot decompose them.

  o However, many of these compounds are toxic and some may be **mutagenic** (mutation causing), **carcinogenic** (cancer causing), or **teratogenic** (birth defect causing).

    ✓ They also may cause liver and kidney dysfunction, sterility, and numerous other physiological and neurological problems.

  o A particularly troublesome class of synthetic organics is the **halogenated hydrocarbons**, organic compounds in which one or more of the hydrogen atoms have been replaced by atoms of chlorine, bromine, fluorine, or iodine.

    ✓ These four elements are classed as halogens; hence the name halogenated hydrocarbons, the commonest being the **chlorinated hydrocarbons** (also called **organic chlorides**).
    ✓ Organic chlorides are widely used in plastics (polyvinyl chloride), pesticides (e.g., Dichlorodiphenyltrichloroethane, which is abbreviated as DDT), solvents (Tetrachlorophenol), electrical insulation (Polychlorinated biphenyls, abbreviated as PCBs) – known to be carcinogenic, mutagenic, and teratogenic, and causes neurological and liver problems.

- **Bioaccumulation**

  o This is accumulation of higher and higher concentrations of potentially toxic chemicals in organisms.
  o It occurs in the case of substances that are ingested but cannot be excreted or broken down (nonbiodegradable substance).

- **Biomagnification**
    - It is the bioaccumulation occurring through several levels of a food chain.

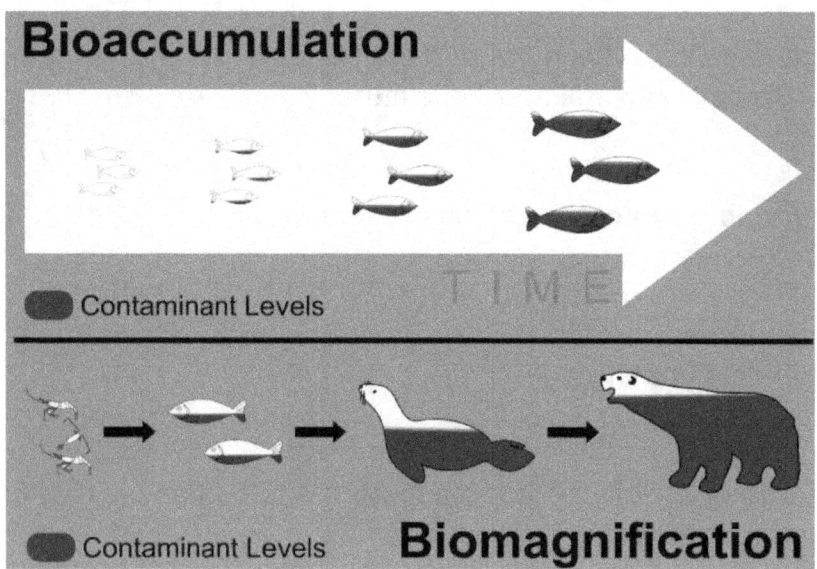

Additional Resources

Internet sites represent a vast resource of information. Using one of the search engines on the Internet such as Google or DuckDuckGo, find more information by searching for words, phrases, or websites: Always use caution when searching for information on the Internet. Follow guidelines to ensure the accuracy and reliability of the information you find.

Search words: HAZMAT, toxic chemicals, bioaccumulate, halogenated hydrocarbons, organic chlorides
Websites: https://www.epa.gov/environmental-topics/chemicals-and-toxics-topics

Self-Assessment

1. List the four hazardous materials and their hazardous properties.

2. Describe the potential consequences of lead and mercury poisoning.

3. Compare mutagenic, carcinogenic, and teratogenic.

4. Provide three examples of bioaccumulation and biomagnification.

25 Major Environmental Laws on Hazardous Wastes

Major Concept

Various agencies and laws control hazardous wastes.

Objectives

- Identify the major environmental laws on hazardous waste
- Describe what each environmental law oversees.

Key Terms

- CAA
- CERCLA
- CERCLA
- CWA
- DOT
- EPA
- EPCRA
- OSHA
- RCRA
- SARA
- SDWA
- Superfund

Chapter Resource

Complementary *full color* illustrations, photos, charts, and graphs are available for Chapter 25 by following this URL: https://tinyurl.com/7vxzdaks This digital resource will enhance your understanding of the chapter concepts.

Environmental Laws

- RCRA - Resource Conservation and Recovery Act

    o The law requires:

    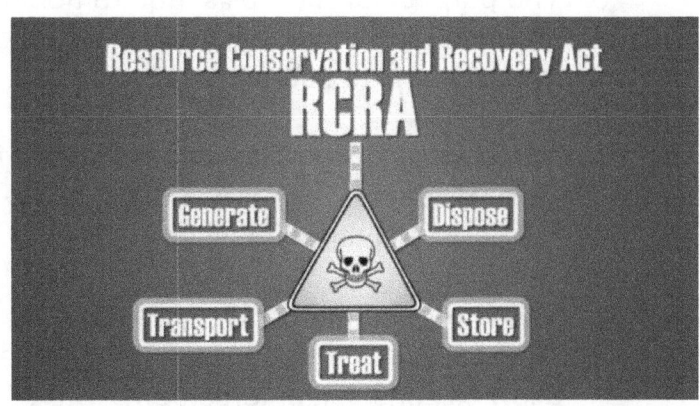

    ✓ Environmental Protection Agency (EPA) to identify hazardous wastes and set standards for their management by states.
    ✓ Firms that store, treat, or dispose of more than 100 kilograms (220 lbs) of hazardous wastes per month to have a permit stating how such wastes are to be managed.
    ✓ Permit holders to use a cradle-to-grave system to keep track of waste transferred from point of origin to approved off-site disposal facilities.

- Superfund

    - This is the **Comprehensive Environmental Response, Compensation, and Liability Act (CERCLA)**.

    - This law has provided a trust fund for:

        - Identifying abandoned hazardous waste dump site and underground tanks leaking toxic chemical
        - Protecting and if necessary, cleaning up ground water near such sites
        - Finding responsible parties to pay for the cleanup of hazardous wastes sites.

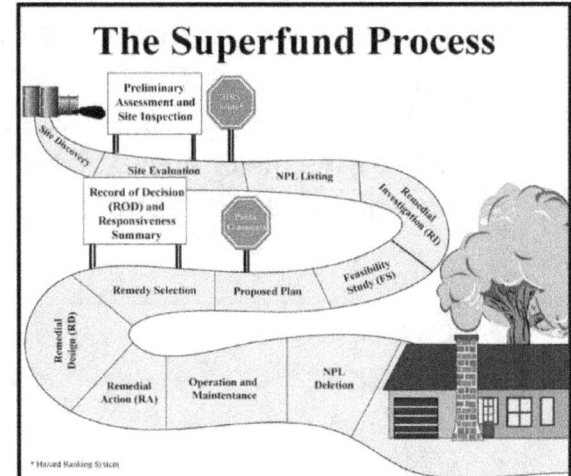

- SARA (Title III)

    - Superfund Amendments and Re-authorization Act – promotes community right to know requirement.

- EPCRA

    - The Federal Emergency Planning and Community Right-to-know Act is adopted as part of Title III of SARA.

    - The purpose of EPCRA's requirements is to promote local emergency planning for chemical releases and to inform citizens of:

        - The existence of toxic chemicals at industrial facilities
        - Spills and other releases

    - These requirements place reporting burdens on industry, provide penalties for noncompliance, and allow citizens suit enforcement.

- TSCA

    - Toxic Substance Control Act – requires new chemicals to be shown safe for specific uses.

- CWA
  - Clean Water Act – limit discharges to waterways.

- SDWA
  - Safe Drinking Water Act – sets standards for drinking water.

- CAA
  - Clean Air Act – limit discharges into the air.

- DOT
  - Department of Transportation Regulations assures safe transportation.

- OSHA
  - Occupational Safety and Health Act – protects workers' health and safety.

Additional Resources

Internet sites represent a vast resource of information. Using one of the search engines on the Internet such as Google or DuckDuckGo, find more information by searching for words, phrases, or websites: Always use caution when searching for information on the Internet. Follow guidelines to ensure the accuracy and reliability of the information you find.

Search words: The names of each of the laws are easily searched by name.

Self-Assessment

1. Identify the standards set for the RCRA law. What must the permit holders provide?

2. What does the Superfund provide funding for?

26 Sustainability Concepts

Major Concept

Sustainability is the balance between the environment, equity, and economy. Ideally it meets the needs of the present without compromising the ability of future generations to meet their own needs.

Objectives

- Define sustainability
- Explain the Three Pillars of Sustainability
- List the seventeen Sustainable Development Goals
- Recognize the principles and concepts of sustainability

Key Terms

- Carrying Capacity
- Consumption
- Etymology
- Organic Movement
- Resiliency

Chapter Resource

Complementary *full color* illustrations, photos, charts, and graphs are available for Chapter 26 by following this URL: https://tinyurl.com/7vxzdaks This digital resource will enhance your understanding of the chapter concepts.

Sustainability Defined

- Achieving sustainability will enable the Earth to continue supporting human life.

- In ecology, sustainability (from sustain and ability) is the property of biological systems to remain diverse and productive indefinitely.

  o Long-lived and healthy wetlands and forests are examples of sustainable biological systems.

- In more general terms, sustainability is the endurance of systems and processes.

  o The organizing principle for sustainability is sustainable development, which includes the four interconnected domains: ecology, economics, politics, and culture.

  o Sustainability science is the study of sustainable development and environmental science.

- Sustainability can also be defined as a socio-ecological process characterized by the pursuit of a common ideal.

  - An ideal is, by definition, unattainable in a given time and space.

  - However, by persistently and dynamically approaching it, the process results in a sustainable system.

- Healthy ecosystems and environments are necessary to the survival of humans and other organisms.

- Ways of reducing negative human impact are environmentally friendly chemical engineering, environmental resources management, and environmental protection.

- Information is gained from green chemistry, earth science, environmental science, and conservation biology.

- Ecological economies study the fields of academic research that aim to address human economies and natural ecosystems.

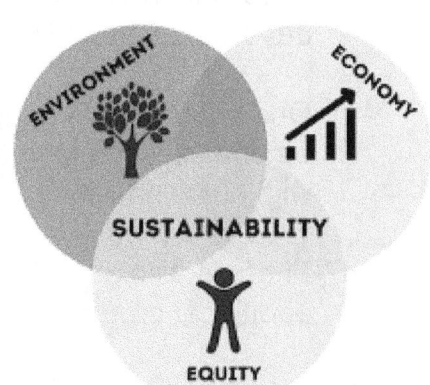

- The term 'sustainability' should be viewed as humanity's target goal of human-ecosystem equilibrium (homeostasis), while "sustainable development" refers to the holistic approach and temporal processes that lead us to the end point of sustainability.

- According to some individuals, despite the increased popularity of the use of the term "sustainability", the possibility that human societies will achieve environmental sustainability has been, and continues to be, questioned – in light of environmental degradation, climate change, overconsumption, population growth and societies' pursuit of unlimited economic growth in a closed system.

Etymology

- The name sustainability is derived from the Latin *sustinere* (tenere, to hold; sub, up).

- Sustain can mean "maintain", "support", or "endure".

- Since the 1980s sustainability has been used more in the sense of human sustainability on planet Earth and this has resulted in the most widely quoted definition of sustainability as a part of the concept sustainable development.

- Sustainable development is development that meets the needs of the present without compromising the ability of future generations to meet their own needs.

Three Pillars of Sustainability

- The 2005 World Summit on Social Development identified sustainable development goals, such as economic development, social development, and environmental protection.

- This view has been expressed as an illustration using three overlapping ellipses indicating that the three pillars of sustainability are not mutually exclusive and can be mutually reinforcing.

- In fact, the three pillars are interdependent, and in the long run none can exist without the others.

- The three pillars have served as a common ground for numerous sustainability standards and certification systems in recent years, in particular, in the food industry.

    o Some sustainability experts and practitioners have illustrated four pillars of sustainability, or a quadruple bottom line.

    o One such pillar is future generations, which emphasizes the long-term thinking associated with sustainability.

    o There is also an opinion that considers resource use and financial sustainability as two additional pillars of sustainability.

- Sustainable development consists of balancing local and global efforts to meet basic human needs without destroying or degrading the natural environment.

    o The question then becomes how to represent the relationship between those needs and the environment.

    o The economy is a subsystem of human society, which is itself a subsystem of the biosphere, and a gain in one sector is a loss from another.

    o This perspective led to the nested circles figure of 'economics' inside 'society' inside the 'environment'.

- The simple definition that sustainability is something that improves "the quality of human life while living within the carrying capacity of supporting eco-systems," though vague, conveys the idea of sustainability having quantifiable limits.

    o But sustainability is also a call to action, a task in progress or "journey" and therefore a political process, so some definitions set out common goals and values.

- Sustainability implies responsible and proactive decision-making and innovation that minimizes negative impact and maintains balance between ecological resilience, economic prosperity, political justice, and cultural vibrancy to ensure a desirable planet for all species now and in the future.

- Specific types of sustainability include:
    - ✓ Sustainable agriculture
    - ✓ Sustainable architecture
    - ✓ Ecological economics

Sustainable agriculture   **Sustainable architecture**

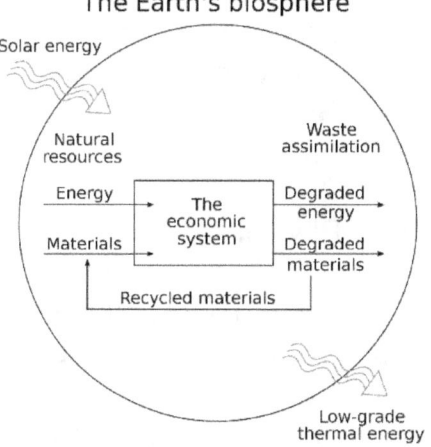

Ecological economics

Circles of Sustainability

- While the United Nations Millennium Declaration identified principles and treaties on sustainable development, including economic development, social development, and environmental protection, it continued using three domains:

    - Economics
    - Environment
    - Social sustainability

- More recently, using a systematic domain model that responds to the debates over the last decade, the Circles of Sustainability approach distinguished four domains:

    - Economic
    - Ecological
    - Political
    - Cultural sustainability

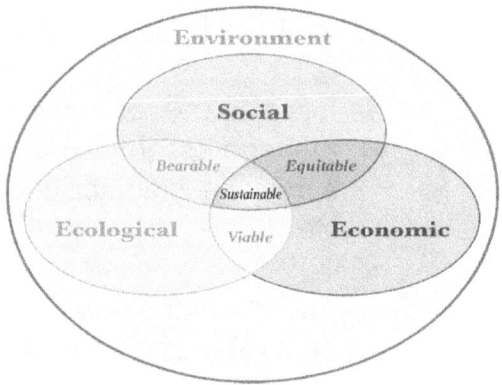

  - ✓ This in accord with the United Nations Agenda 21, which specifies culture as the fourth domain of sustainable development.

Seven Modalities

- Another model suggests humans attempt to achieve all of their needs and aspirations via seven modalities:

    - Economy
    - Community
    - Occupational groups
    - Government
    - Environment
    - Culture
    - Physiology

- From the global to the individual human scale, each of the seven modalities can be viewed across seven hierarchical levels.

- Human sustainability can be achieved by attaining sustainability in all levels of the seven modalities.

Resiliency

- Resiliency in ecology is the capacity of an ecosystem to absorb disturbance and still retain its basic structure and viability.

- Resilience-thinking evolved from the need to manage interactions between human-constructed systems and natural ecosystems in a sustainable way despite the fact that to policymakers a definition remains elusive.

- Resilience-thinking addresses how much planetary ecological systems can withstand assault from human disturbances and still deliver the services current and future generations need from them.

    - It is also concerned with commitment from geopolitical policymakers to promote and manage essential planetary ecological resources in order to promote resilience and achieve sustainability of these essential resources for benefit of future generations of life.

    - The resiliency of an ecosystem, and thereby, its sustainability, can be reasonably measured at stages or events where the combination of naturally occurring regenerative forces (solar energy, water,

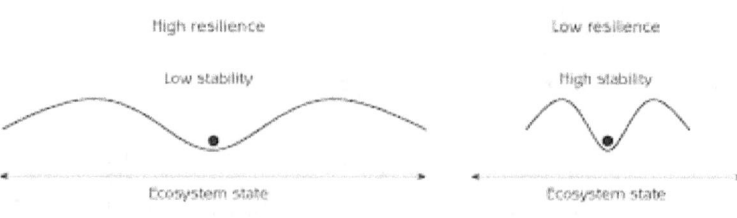

soil, atmosphere, vegetation, and biomass) interact with the energy released into the ecosystem from disturbances.

- A practical view of sustainability is closed systems that maintain processes of productivity indefinitely by replacing resources used by actions of people with resources of equal or greater value by those same people without degrading or endangering natural biotic systems.

    o In this way, sustainability can be concretely measured in human projects if there is a transparent accounting of the resources put back into the ecosystem to replace those displaced.

- In nature, the accounting occurs naturally through a process of adaptation as an ecosystem returns to viability from an external disturbance.

    o The adaptation is a multi-stage process that begins with the disturbance event (earthquake, volcanic eruption, hurricane, tornado, flood, or thunderstorm), followed by absorption, utilization, or deflection of the energy or energies that the external forces created.

- In analyzing systems such as urban and national parks, dams, farms and gardens, theme parks, open-pit mines, and water catchments, one way to look at the relationship between sustainability and resiliency is to view the former with a long-term vision and resiliency as the capacity of human engineers to respond to immediate environmental events.

History

- The history of sustainability traces human-dominated ecological systems from the earliest civilizations to the present time.

- This history is characterized by the increased regional success of a particular society, followed by crises that were either resolved, producing sustainability, or not, leading to decline.

- In early human history, the use of fire and desire for specific foods may have altered the natural composition of plant and animal communities.

    o Between 8,000 and 10,000 years ago, agrarian communities emerged which depended largely on their environment and the creation of a "structure of permanence."

- The Western industrial revolution of the 18th to 19th centuries tapped into the vast growth potential of the energy in fossil fuels.

- Coal was used to power ever more efficient engines and later to generate electricity.

- Modern sanitation systems and advances in medicine protected large populations from disease.

- In the mid-20th century, a gathering environmental movement pointed out that there were environmental costs associated with the many material benefits that were now being enjoyed.

- In the late 20th century, environmental problems became global in scale.

- The 1973 and 1979 energy crises demonstrated the extent to which the global community had become dependent on non-renewable energy resources.

[Diagram: Arrows from Agriculture crops & livestock, Urbanisation, Industry, Forestry, Fisheries, Nature, Tourism, and Domestic water supply all pointing toward "Competition for land and water resources" in the center.]

- The current thinking in the 21st century: There is increasing global awareness of the threat posed by the human greenhouse effect, produced largely by forest clearing and the burning of fossil fuels.

Principles, Scales, and Context

- The philosophical and analytic framework of sustainability draws on and connects with many different disciplines and fields.

  - In recent years, an area that has come to be called sustainability science has emerged.

- Sustainability is studied and managed over many scales (levels or frames of reference) of time and space and in many contexts of environmental, social, and economic organization.

- The focus ranges from the total carrying capacity (sustainability) of planet Earth to the sustainability of economic sectors, ecosystems, countries, municipalities, neighborhoods, home gardens, individual lives, individual goods and services, occupations, lifestyles, behavior patterns, and so on.

  - In short, it can entail the full compass of biological and human activity of any part of it.

  - The sheer size and complexity of the planetary ecosystem has proved problematic for the design of practical measure to reach global sustainability.

Consumption

- A major driver of human impact on Earth systems is the destruction of biophysical resources, and especially, the Earth's ecosystems.

- The environmental impact of a community or of humankind as a whole depends both on population and impact per person, which in turn depends in complex ways on what resources are being used, whether or not those resources are renewable, and the scale of the human activity relative to the carrying capacity of the ecosystems involved.

- Careful resource management can be applied at many scales, from economic sectors like agriculture, manufacturing, and industry, to work organizations, the consumption patterns of households and individuals and to the resource demands of individual goods and services.

- One of the initial attempts to express human impact mathematically was developed in the 1970s and is called the I PAT formula.

    o This formulation attempts to explain human consumption in terms of three components:

    - Population numbers
    - Levels of consumption (which it terms "affluence", although the usage is different)
    - Impact per unit of resource used (which is termed "technology", because this impact depends on the technology used).

    o The equation is expressed:

    $I = P \times A \times T$

    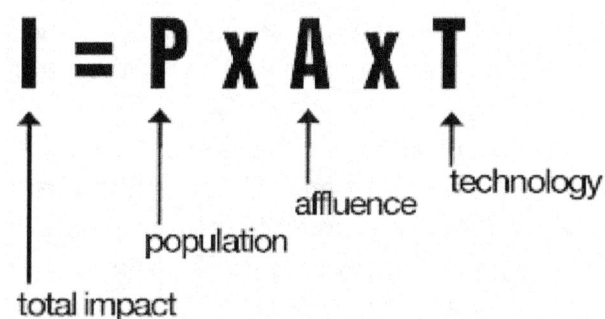

    Where I = Environmental impact, P = Population, A = Affluence, T = Technology

Measurement

- Sustainability measurement is a term that denotes the measurements used as the quantitative basis for the informed management of sustainability.

- The metrics used for the measurement of sustainability (involving the sustainability of environmental, social, and economic domains, both individually and in various combinations) are evolving.

    o They include indicators, benchmarks, audits, sustainability standards and certification systems like Fairtrade and Organic, indexes and accounting, as well as assessment, appraisal, and other reporting systems.
    o They are applied over a wide range of spatial and temporal scales.

13 Standards of Sustainable Agriculture

- These standards were developed by National Agricultural Institute (national-ag-institue.org)

    1. Bases directions and changes on science
    2. Honors market principles
    3. Increases profitability and reduce risks
    4. Satisfies human need for fiber and safe, nutritious food
    5. Conserves and seeks energy resources
    6. Creates and conserves healthy soil
    7. Conserves and protects water resources
    8. Recycles or manages waste products
    9. Selects animals and crops appropriate for the environment and available resources
    10. Manages pests with minimal environmental impact
    11. Encourages strong rural communities
    12. Promotes social and environmental responsibilities
    13. Uses appropriate technology

Population Growth

- According to the most recent (2019) revision of the official United Nations World Population Prospects, the world population is projected to reach 8.5 billion by 2030, 9.7 billion people by 2050, and to reach 10.9 billion by the year 2100.

    - With a projected addition of over one billion people, countries of sub-Saharan Africa could account for more than half of the growth of the world's population.

    - In 2018, for the first time in history, persons aged 65 years or over worldwide outnumbered children under age five. Projections indicate that by 2050 there will be more than twice as many persons above 65 as children under five. By 2050, the number of persons aged 65 years or over globally will also surpass the number of adolescents and youth aged 15 to 24 years.

- Life expectancy at birth for the world's population reached 72.6 years in 2019, an improvement of more than 8 years since 1990. Further improvements in survival are projected to result in an average length of life globally of around 77.1 years in 2050.

    - It is estimated that ten countries are experiencing a net outflow of more than 1 million migrants between 2010 and 2020. For many of these, losses of population due to migration are dominated by temporary labour movements, such as for Bangladesh (net outflow of -4.2 million during 2010-2020), Nepal (-1.8 million) and the Philippines (-1.2 million).

    - In others, including Syria (-7.5 million), Venezuela (-3.7 million), and Myanmar (-1.3 million), insecurity and conflict have driven the net outflow of migrants over the decade.

Carrying Capacity

- Ecological footprint for different nations compared to their Human Development Index (HDI).

- At the global scale, scientific data now suggests that humans are living beyond the carrying capacity of planet Earth.

- The ecological footprint measures human consumption in terms of the biologically productive land needed to provide the resources, and absorb the wastes of the average global citizen.

    o In 2008 it required 2.7 global hectares per person, 30% more than the natural biological capacity of 2.1 global hectares (assuming no provision for other organisms).

    o The resulting ecological deficit must be met from unsustainable extra sources and these are obtained in three ways:

        ✓ Embedded in the goods and services of world trade
        ✓ Taken from the past (e.g., fossil fuels)
        ✓ Borrowed from the future as unsustainable resource usage (e.g., by over exploiting forests and fisheries)

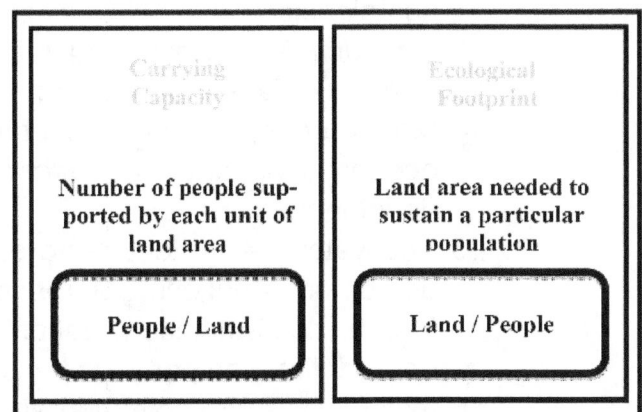

- At a fundamental level, energy flow and biogeochemical cycling set an upper limit on the number and mass of organisms in an ecosystem.

- Human impacts on the Earth are demonstrated in a general way through detrimental changes in the global biogeochemical cycles of chemicals that are critical to life, most notably those of water, oxygen, carbon, nitrogen, and phosphorus.

Sustainable Development Goals

- The Sustainable Development Goals (SDGs) are the current harmonized set of 17 future international development targets.

- The Official Agenda for Sustainable Development adopted on 25 September 2015 has 92 paragraphs, with the main paragraph (51) outlining the 17 Sustainable Development Goals and its associated 169 targets.
    o This includes the following 17 goals:

    1. Poverty – End poverty in all its forms everywhere
    2. Food – End hunger, achieve food security and improved nutrition and promote sustainable agriculture
    3. Health – Ensure healthy lives and promote well-being for all at all ages
    4. Education – Ensure inclusive and equitable quality education and promote lifelong learning opportunities for all
    5. Women – Achieve gender equality and empower all women and girls

6. Water – Ensure availability and sustainable management of water and sanitation for all
7. Energy – Ensure access to affordable, reliable, sustainable, and modern energy for all
8. Economy – Promote sustained, inclusive, and sustainable economic growth, full and productive employment, and decent work for all
9. Infrastructure – Build resilient infrastructure, promote inclusive and sustainable industrialization and foster innovation
10. Inequality – Reduce inequality within and among countries
11. Habitation – Make cities and human settlements inclusive, safe, resilient, and sustainable
12. Consumption – Ensure sustainable consumption and production patterns
13. Climate – Take urgent action to combat climate change and its impacts
14. Marine-ecosystems – Conserve and sustainably use the oceans, seas, and marine resources for sustainable development
15. Ecosystems – Protect, restore, and promote sustainable use of terrestrial ecosystems, sustainably manage forests, combat desertification, and halt the reverse land degradation and halt biodiversity loss
16. Institutions – Promote peaceful and inclusive societies for sustainable development, provide access to justice for all and build effective, accountable, and inclusive institutions at all levels
17. Sustainability – Strengthen the means of implementation and revitalize the global partnership for sustainable development

Applied Sustainability

- Applied sustainability is the application of science and innovation, including the insights of the social sciences, to meet human needs while indefinitely preserving the life support systems of the planet.

- A more refined definition would be called just applied sustainability: the just and equitable application of science and innovation, including the

social sciences, to ensure a better quality of life for all, now and into the future whilst living within the limits of supporting ecosystems.

- o It generates an improved definition of sustainable development as *"the need to ensure a better quality of life for all, now and into the future, in a just and equitable manner, whilst living within the limits of supporting ecosystems."*

- o This new form of sustainable development prioritizes justice and equity, while maintaining the importance of the environment and the global life support system.

- The relationship between applied sustainability and sustainability (or sustainability science) is an analogous relationship between applied science (engineering) and basic science.

  - o Whereas science is the effort to discover, understand, or to understand better, how the physical world works, with observable physical evidence as the basis of that understanding.

  - o Applied science is the application of knowledge from one or more natural scientific fields to solve practical problems.

Sustainable Engineering

- Applied sustainability is essentially sustainable engineering.

- By using natural laws and physical resource in order to design and implement materials, structures, machines, devices, systems, and processes that meets human need while preserving the environment forever.

- Applied sustainability is made up of work in engineering, policy, and education – whatever methods are necessary to conserve the world for our children.

- A recent study has shown that open-source principles can be used to accelerate deployment of sustainable technologies such as open source appropriate technologies.

Sustainable Counterpoint by Henry I. Miller (2017)

- Henry I. Miller, a physician and molecular biologist, is the Robert Wesson Fellow in Scientific Philosophy and Public Policy at Stanford University's Hoover Institution.

  - o He was the founding director of the FDA's Office of Biotechnology.

- 'Sustainable' has become a buzzword applicable not only to agriculture and energy production but to sectors as far afield as the building and textile industries.

- Many large companies tout the concept and boast a sustainability department, and the United Nations has hundreds of projects concerned with sustainability throughout its many agencies and programs.

  - Some universities offer courses or even degrees in "sustainability".

  - One of Stanford University's oldest dormitories, Roble Hall, houses an initiative called the Roble Living Laboratory for Sustainability at Stanford (ROLLSS), which includes "undergraduate seminars, a graduate-student speaker series, and activities intended to engage the dorm's residents in curbing their natural-resource waste."

  - So far, so good, but a central part of the initiative is an organic garden – which is not so good because the students are being schooled in the myth that organic agricultural methods are sustainable and ethical.

    - ✓ And that sophistry (a fallacious argument) is by no means limited to one dormitory; all eight of Stanford's major dining halls maintain an organic "dedicated teaching garden."

The Organic Movement

- Advocates of organic agriculture tout is as a "sustainable" way to feed the planet's expanding population.

- According to the Worldwatch Institute, *"Organic farming has the potential to contribute to sustainable food security by improving nutrition intake and sustaining livelihoods in rural areas, while simultaneously reducing vulnerability to climate change and enhancing biodiversity."*

  - This is wishful thinking, if not outright delusion, and unfortunately, it is being promulgated on elite university campuses, including Stanford's.

- The organic movement touts the sustainability of its methods, but the claims do not withstand scrutiny.

  - For example, a study published in Hydrology and Earth System Sciences found that the potential for groundwater contamination can be dramatically reduced if fertilizers are distributed through the irrigation system according to plant demand during the growing season.

- But organic farming depends on compost, the release of which is not matched with plant demand.

    - The study found that "intensive organic agriculture relying on solid organic matter, such as composted manure that is implemented in the soil prior to planting as the sole fertilizer, resulted in significant down-leaching of nitrate" into groundwater.
    - With many of the world's most fertile farming regions in the throes of drought and aquifer depletion – which was the subject of a 60 Minutes segment – increased nitrate in groundwater is hardly a mark of sustainability.

    - Moreover, although composting gets good PR as a "green" activity, on a large scale it generates a significant amount of greenhouse gases (and is also often a source of pathogenic bacteria applied to crops).

- Organic farming might work well for certain local environments on a small scale, but it is hugely wasteful of arable land and water because of its low yields.

    - Organic farms produce far less food per unit of land and water than conventional ones.

    - Plant pathologist Dr. Steve Savage analyzed the data from the U.S. Department of Agriculture's 2014 Organic Survey, which reports various measures of productivity from most of the certified organic farms in the nation, and compared they to those at conventional farms, crop by crop, state by state. His findings are extraordinary.

        - ✓ Of the 68 crops surveyed, there was a "yield gap" – poorer performance of organic farms – in 59.
        - ✓ And many of those gaps, or shortfalls, were impressive: strawberries, 61% less than conventional; fresh tomatoes, 61% less, tangerines, 58% less; carrots, 49% less; cotton, 45% less; rice, 39% less; peanuts, 37% less.
        - ✓ These findings are important, as Savage observed.

    - To have raised all U.S. crops as organic in 2014 would have required farming of 109 million more acres of land.

        - ✓ That is an area equivalent to all the parkland and wildland areas in the lower 48 states, or 1.8 times as much as all the urban land in the nation.

- The low yields of organic agriculture impose a variety of stresses on farmland and especially on water consumption.

- A British meta-analysis published in the Journal of Environmental Management (2012) addressed the question whether organic farming reduces environmental impacts.

- - It identified some of the stresses that were higher in organic, as opposed to conventional, agriculture: "ammonia emissions, nitrogen leaching, and nitrous oxide emissions per product unit were higher from organic systems," as were "land use, eutrophication potential and acidification potential per product unit."

- Lower organic crop yields are largely inevitable, given the arbitrary rejection of various advanced methods and technologies.

- Organic agriculture affords limited pesticide options, difficulties in meeting peak fertilizer demand, and a lack of access to varieties modified with the most precise and predictable techniques of genetic engineering.

  - If the scale of organic production were significantly increased, the lower yields would increase the pressure for the conversion of more land to farming and the burden on water supplies, both of which are serious environmental issues.

  - In short, organic practices are to agriculture what cigarette smoking is to human health.

- The issue of water conservation, in particular, illustrates an irony in the Stanford ROLLSS program.

  - Consider this from a recent Stanford publication: *"Two freshmen particularly active this year [in the ROLLSS program], Raja Ramesh and Kyle Enriquez, have taken it upon themselves to run initiatives encouraging students to conserve water by washing fuller loads of laundry and by taking shorter showers."* Why don't they, in addition, adopt modern, water-conserving non-organic farming practices.

- Organic production disfavors the best approach to enhancing soil quality – the minimization of soil disturbances (e.g., no plowing or tilling), combined with the use of cover crops.

  - Such farming systems offer multiple environmental advantages, particularly with respect to limited erosion, the runoff of fertilizers and pesticides, and the release of CO_2 from tilling.

  - Organic growers do frequently plant cover crops, but in the absence of effective herbicides, they often rely on tillage (or even labor-intensive hand-weeding) for weed control.

- Many who are seduced by the romance of organic farming (read: college students) ignore its human consequences.

- - American farmer Blake Hurst offers this reminder: *"Weeds continue to grow, even in polycultures with holistic farming methods, and, without pesticides, hand weeding is the only way to protect a crop."*

  - The back-breaking drudgery of hand weeding often falls to women and children.

- One prevalent "green myth" about organic agriculture is that it does not employ pesticides.

- Organic farming does, in fact, use insecticides and fungicides to prevent predation of its crops.

  - More than 20 chemicals (mostly containing copper and sulfur) are commonly used in the growing and processing of organic crops and are acceptable under USDA's arbitrary organic rules.

- The exclusion of certain organisms from organic agriculture simply because they were crafted with superior molecular techniques makes no sense.

- Perhaps the most illogical and least sustainable aspect of organic farming in the long term will turn out to be the systematic and absolute exclusion of "genetically engineered" plants – but only those that were modified with the most precise and predictable modern molecular techniques.

  - Except for wild berries and wild mushrooms, virtually all the fruits, vegetables, and grains in our diet have been genetically improved by one technique or another – often as a result of seeds having been irradiated or via wide crosses, which move genes from one species or genus to another in ways that do not occur in nature.

- In recent decades, we have seen advances in agriculture, such as plants that are drought- or flood-resistant, that make farming more environmentally friendly and sustainable than ever before.

  - But they have resulted from science-based research and technological ingenuity on the part of farmers, plant breeders, and agribusiness companies, not from ignorant, arrogant social elites disdainful of modern insecticides, herbicides, genetic engineering, and "industrial agriculture."

  - And here is another cosmic irony: The co-discoverer in 1973 of recombinant DNA technology – the prototypic, iconic molecular technique for genetic engineering – was Stanford biochemist Dr. Stanley N. Cohen, who is still a professor of genetics and medicine at the university.

- ✓ I wonder how many of the New Age sustainability advocates involved in the ROLLSS program have even heard of him.

- As genetic engineering's successes continue to emerge, the gap between modern, high-tech agriculture and organic methods will become a chasm.

- Genetically engineered, drought-resistant, flood-resistant and "fortified" crops have begun to emerge from the development pipeline, and genetically engineered potato varieties in the marketplace are bruise-resistant and contain 50 to 70% less asparagine, a chemical that is converted to acrylamide, a probable carcinogen, when heated to high temperatures.

  - The advantage of lower levels of acrylamide is obvious, but the bruise-resistance is important to sustainability.

  - According to Simplot, the developer of the genetically engineered "Innate" varieties, *"with full market penetration for its varieties sold in the U.S., Innate will reduce annual potato waste by an estimated 400 million pounds in the food service and retail industries and a significant portion of the estimated 3 billion pounds discarded by consumers."*

    - ✓ And a second generation of Innate potatoes now completing regulatory review contains an additional trait: resistance to a destructive fungus called "late blight," which caused the Irish potato famine of the mid-19th century and is still with us.

  - Potatoes that resist bruising and the late blight are major advances in sustainability because every serving of French fries or mashed potatoes made from them requires less farmland and less water.

  - But none of those varieties can be used by organic farmers, including the Stanford students at Roble Hall.

    - ✓ The ROLLSS program claims to have "drawn support and involvement from institutions across Stanford," and lists them, but he entities that contain the university's eminent genetic engineers and plant scientists do not seem to be participating.

    - ✓ One has to wonder how one of the world's preeminent research universities, which regularly produces breakthroughs across the entire spectrum of science, technology, and engineering, could so blindly embrace and endorse destructive practices worthy of the 19th century.

- In an article titled "The Organic Fable," New York Times columnist Roger Cohen had some pithy observations about the popularity of organic food, including this one:

- *"Organic has long become an ideology, the romantic back-to-nature obsession of an upper middle class able to afford it and oblivious, in their affluent narcissism, to the challenge of feeding a planet whose population will surge to nine billion before the middle of the century and whose poor will get a lot more nutrients from the two regular carrots they can buy for the price of one organic carrot."*

- Here is a suggestion for a new tradition at Roble Hall: Ditch organic agriculture, begin taking advantage of modern technologies to boost yields, and commemorate that decision each year with an event called "Two Carrot Day." (It could be co-sponsored by Stanford's Center for Compassion and Altruism Research and Education.)

- Sustainability has been defined this way:

  - The ability to provide for the needs of the current population without damaging the ability of future generations to provide for themselves.

  - When a process is sustainable, it can be carried out over and over without negative environmental effects or impossibly high costs to anyone involved.

  - That definition is compatible with the notion that sustainable farming is favored by maximizing human ingenuity and the quest for progress – that is, for inventing processes and products that are more efficient, less costly, and at the same time, less harmful to the environment.

    - ✓ In other words, exactly the kinds of things that come from universities' chemistry, plant-science, and molecular-biology labs.
    - ✓ But organic farmers, including Stanford's, can forget about using them.

Additional Resources

Internet sites represent a vast resource of information. Using one of the search engines on the Internet such as Google or DuckDuckGo, find more information by searching for words, phrases, or websites: Always use caution when searching for information on the Internet. Follow guidelines to ensure the accuracy and reliability of the information you find.

Search words: Circles of Sustainability, seven modalities, ecological resiliency, carrying capacity, global human impact on biodiversity, sustainable engineering, organic movement

Website: https://population.un.org/wpp/Publications/Files/WPP2019_Highlights.pdf

Self-Assessment

1. Define sustainable development.

2. Identify what drives the competition for land and water resources.

3. Compare organic agriculture to conventional agriculture.

4. List three examples of sustainable engineering.

27 Basics of Composting

Major Concept

Composting is the controlled biological decomposition and conversion of solid organic material into a humus like substance called compost.

Objectives

- Describe the process of composting
- List five major parameters of composting efficiently
- Provide the correct ratios of raw materials for composting
- Name the primary nutrients required by the microorganisms involved in composting

Key Terms

- Aerobic
- Anaerobic
- C:N ratio
- Composting
- Mesophilic
- Thermophilic

Chapter Resource

Complementary *full color* illustrations, photos, charts, and graphs are available for Chapter 27 by following this URL: https://tinyurl.com/7vxzdaks This digital resource will enhance your understanding of the chapter concepts.

Composting

- Composting is the controlled biological decomposition and conversion of solid organic material into a humus like substance called compost.

- Composting is the process of letting nature transform organic materials into a material with environmentally beneficial applications.

- The process is aerobic, meaning it requires oxygen.

- The process uses various microorganisms such as bacteria, actinomyces and fungi to break down the organic compounds into simpler substances.

    o In natural surroundings, leaves and branches that fall to the ground form a rich, moist layer of mulch that protects the roots of plants and provides a home for nature's most fundamental recyclers: worms, insects, and a host of microorganisms too small to see with the naked eye.

    o Composting is a viable process of treating solid waste for beneficial use and destroying pathogens, diseases, and undesirable weed seed.

- By responsibly managing air, moisture and nutrients, the composting process can transform large quantities of organic material into compost in a relatively short time.

- During composting, the microorganisms consume oxygen while feeding on organic matter.

- Active composting generates a considerable amount of heat, and large quantities of carbon dioxide and water vapor are released into the air.

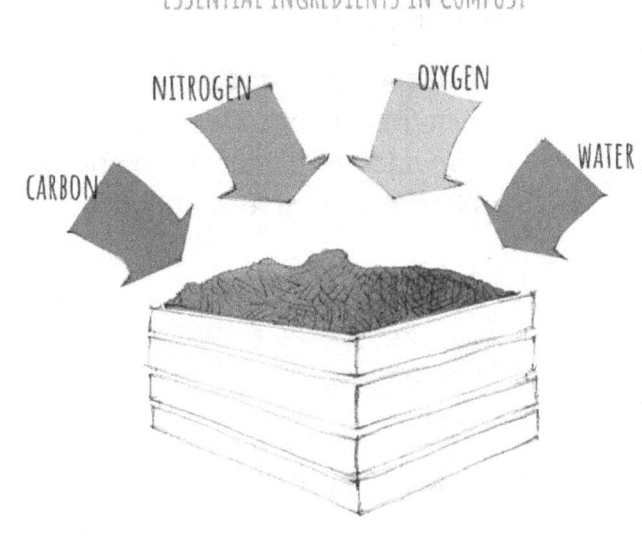

  - The carbon dioxide and water losses can amount to half the weight of the initial organic materials, so composting reduces both the volume and mass of the raw materials while transforming them into a beneficial humus-like material.

- Composting is most efficient when the major parameters – oxygen, nitrogen, carbon, moisture, and temperature – which affect the composting process, are effectively managed.

Oxygen and Aeration

- An aerobic composting process consumes large amounts of oxygen.

- During the first days of composting, easily degradable components of the organic materials are rapidly metabolized.

- The need for oxygen and the production of heat are greatest during the early stages and then decrease as the process continues.

  - If the supply of oxygen is limited, the composting process slows and the process becomes anaerobic (without oxygen).
  - A minimum oxygen concentration of 5% within the pore spaces of the composting material is recommended for a well-managed compost facility (air contains about 21% oxygen).

Temperature

- In addition to providing oxygen, aeration removes heat, water vapor and other gases trapped within the composting materials.

  - In fact, the required rate of aeration for heat removal can be 10 times greater than for supplying oxygen.

- Temperature frequently determines how much and how often aeration is required.

- As a matter of convenience, science has subdivided and given names to the temperature ranges in which certain microorganisms are most active.

- Composting is most efficient when the temperature of the composting material is within the two ranges known as Mesophilic (80°-120° F) and Thermophilic (105°-150° F).

    - Mesophilic temperatures allow effective composting, but most experts suggest maintaining temperatures between 110° and 150° F.

    - The Thermophilic, or higher, composting temperatures are desirable because they destroy more pathogens, diseases, weed seeds and insect larvae in the composting materials.

- The regulations have set the critical temperature for killing human pathogens at 55°C (131°F) for a specified time.

    - This time and temperature should destroy most plant pathogens as well.

    - The critical temperature for destroying most weed seed is 145°F.

- Microbial decomposition during composting releases large amounts of energy as heat.

- The insulating qualities of the composting materials lead to an accumulation of heat, which raises the temperature.

- The materials lose head as water evaporates and as air movement carries away the water vapor and other gases.

    - Turning and aeration of a compost pile accelerates the heat loss and can be used to reduce the temperature.

- Heat accumulation in a compost pile can rise about 160°F (71°C) because of the heat generated by microbial activity and the insulating qualities of the composting materials.

    - When the temperature reaches this level, many of the microbes die or become dormant.

    - The composting process effectively stops and does not recover until the population of the microorganisms recovers.

    - The temperatures should be monitored and, when the composting material becomes too hot, heat loss should be accelerated by forced aeration or by turning.

    - Since most of the heat loss from composting occurs by the evaporation of water, the materials should not be allowed to dry below 40% moisture.

- Low moisture increases the chance of high temperature damage and spontaneous combustion.

Moisture

- Moisture is the lifeblood of the metabolic processes of the microbes.

- Water provides the medium for chemical reactions, transport nutrients and allows the microorganisms to move from place to place.

- In theory, biological activity is optimum when the materials are saturated.

    - Activity ceases entirely below a 15% moisture content.
    - Efficient activity is achieved when the moisture is maintained between 40% and 60%.
    - At moisture levels above 60%, water displaces much of the air in the pore spaces of the composting materials.
        - ✓ This limits air movement and leads to anaerobic conditions.
    - The desired starting moisture content of organic materials composted should be higher than 40%.

- The moisture content generally decreases as the organic material decomposes.

- For co-composting, mix materials that are too wet with materials that are too dry to achieve a 50%-60% moisture content of the blended materials.

- With dry materials, such as leaves, add water directly to the compost pile.

    - Generally, more moisture will evaporate than is added to the composting windrow by rain.

- Moisture levels should be maintained so the materials are thoroughly wet but not waterlogged or dripping excessive water.

    - There are instruments and procedures to accurately determining moisture content, but as a rule, the materials are too wet if water can be squeezed out of a handful and too dry if the handful does not feel moist to the touch.

- The optimum time to add water is during the aeration or turning operation.

Odors

- Odors are the single biggest threat to a composting operation.

- Nothing is more persistent than an angry neighbor capable of causing a composting operation to close because of odors.

- Theoretically, aerobic composting does not generate odorous compounds as the anaerobic process does.

- Objectionable odors can come from certain raw materials or the process itself if composting is not responsibly managed.

- There are three primary sources of odors at a composting facility:

    1. Odorous raw materials
    2. Ammonia lost from high nitrogen materials.
    3. Anaerobic conditions within the windrows or composting facility

- Anaerobic conditions can be minimized by proper management.

- Use a good mix of raw materials, avoid overly wet mixes, manage leachate and rainfall runoff, monitor temperatures, and turn to aerate the materials regularly.

- Bad odors can be controlled by providing extra carbon in the mix and maintaining the pH below 8.5.

    o A high pH encourages the conversion of nitrogen compounds to ammonia, which further adds to alkalinity.

- The most common causes of odors at a composting site are strong-smelling raw materials.

    o Odors come to the site with the materials but, with proper management, should dissipate when the materials begin composting.

Nutrients

- Carbon (C), nitrogen (N), phosphorus (P), and potassium (K) are the primary nutrients required by the microorganisms involved in composting.

    o Nitrogen, phosphorus, and potassium are also the primary nutrients for plants, so nutrient concentrations also influence the value of the compost.

- Many organic materials contain enough quantities of nutrients for composting.

- - Excessive or insufficient carbon or nitrogen will affect the process.
  - Carbon provides microorganisms with both energy and growth; nitrogen is essential for protein and reproduction.

- In general, biological organisms need about 25 times more carbon than nitrogen.

- Carbon and nitrogen need to be provided appropriate proportions.

  - The ratio of carbon to nitrogen is referred to as the C:N ration.
  - Raw materials blended to provide a C:N ratio of 25:1 to 30:1 are ideal for active composting, although initial ratios of 20:1 up to 40:1 consistently give good results.
  - For some applications, C:N ratios of 50:1 and higher are acceptable.
  - With C:N ratios below 20:1 the available carbon is fully used without stabilizing all of the nitrogen.
  - The excess nitrogen may be lost to the atmosphere as ammonia or nitrous oxide, and odor can become a problem.
  - Mixes with C:N ratios higher than 40:1 require longer composting times for the microorganisms to use the excess carbon because of the absence of available nitrogen.

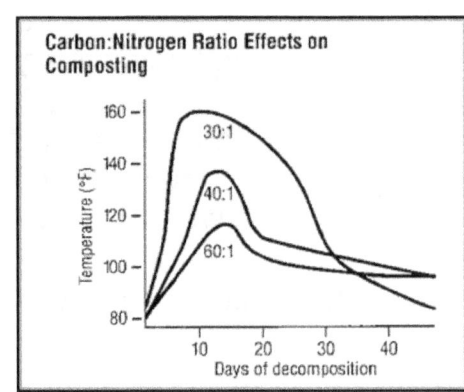

- The C:N ratio is a useful guide to formulate composting recipes, but the rate at which the carbon compounds decompose must also be considered.

  - For example, straw decomposes and releases carbon to the microorganisms more easily than woody materials.
  - This occurs because the carbon compounds in woody materials are bound by lignins which are organic compounds highly resistant to biological breakdown.
  - Similarly, carbon in the simple sugars of fruit wastes is more quickly consumed by the microorganisms than the cellulose carbon in straw.

- Decomposition by the microorganisms occurs on the surface of the organic particle.

  - Therefore, degradability can be improved by reducing the particle size, which increases the surface area, as long as porosity does not become a problem by decreasing the airspace and needed oxygen.

- Other considerations include pH, time, and co-composting.

- pH
    - Composting process is relatively insensitive to pH within the range commonly found in mixtures of organic materials, largely because of the broad spectrum of microorganisms involved.
    - Preferred pH is in the range of 6.5-8.0.
    - pH does become a consideration with raw materials containing a high percentage of nitrogen.
    - A high pH, above 8.5, encourages the conversion of nitrogen compounds to ammonia.

- Time
    - The time required to transform raw materials into compost depends on many factors.
    - Proper moisture content, C:N ratio, and frequent aeration ensure the shortest composting period.
    - A well-managed composting operation should produce quality compost within four months.

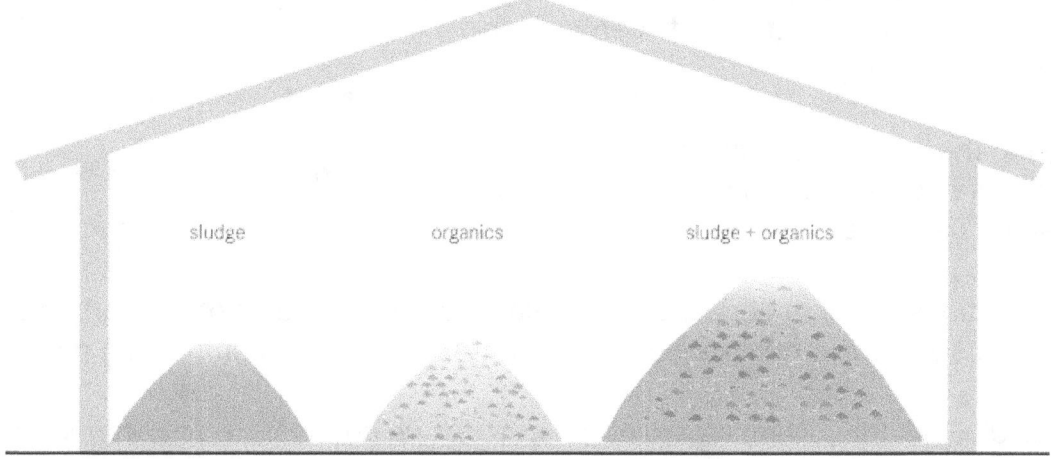

- Co-composting
    - Co-composting refers to composting that includes more than one organic material.
    - Some co-composting operations determine the portions of each material by trial and error to obtain a compostable mixture. The trial-and-error method could cause problems later.
    - To obtain the best ingredients for composting within the optimum time period without excessive odors, follow a mixing procedure based on the physical and chemical characteristics of the composting materials.
    - When the material characteristics are unknown, the look and feel method may be the only alternative.

Math of Composting

- Developing a recipe for composting can sometimes be a balancing act to get the C:N ratio as well as the moisture content within acceptable ranges.
- Moisture content can be critical. Excessive moisture will lead to anaerobic conditions, odors and will slow decomposition.

    o The effect of C:N ratios less than ideal is less critical.
    o It's best to develop the initial recipe based on moisture content and adjust the portions of materials to obtain an acceptable C:N ratio.

- Formulas for calculating a composting recipe are presented. To use the equations, you must know the carbon and nitrogen characteristics of the materials.

- If the carbon and nitrogen contents of a material are not known, it's not unusual to select a material similar in characteristics with known C:N values and calculate C:N ratios for the desired recipe of materials.

- The ratio of mixed materials is adjusted until a desired moisture content between 40% to 60% is achieved.

- Using the selected materials based on moisture content, the C:N ratio of the mixture can be calculated.

$$\% \text{ Moisture Content} = \frac{\text{Wet Weight} - \text{Dry Weight}}{\text{Wet Weight}} \times 100$$

$$\text{Moisture Content (wt.)} = \frac{\% \text{ Moisture}}{100}$$

$$\text{Dry Weight} = \text{Total weight} - \text{Weight of Water}$$

- Proportions of dry materials can be calculated on the basis of the carbon and nitrogen contents since it is relatively easy to add moisture to the composting mix.

- Calculations for C:N ratios are done on a dry weight basis, so it's important to know how the laboratory results are determined, on a wet or dry basis.

$$\text{Nitrogen Content} = \text{Dry Weight} \times (\%N/100)$$

$$\text{Carbon Content} = \text{Dry Weight} \times (\%C/100)$$

$$\text{Carbon Nitrogen Ratio (C/N)} = \frac{\text{Carbon Content}}{\text{Nitrogen Content}}$$

Equations for mixing materials:

(1) $\text{Mix Moisture} = \dfrac{\text{Wt. of Water in material A + water in B + water in C+...}}{\text{Total Weight of all Material}}$

$$\text{Weight of water to be added, } M^x = \frac{W(M^d - M^i)}{(1 - M^d)}$$

Symbols:

M^x = Amount of water to be added, tons
M^d = Desired moisture content, % total weight/100
M^i = Initial material moisture content/100
W = Initial feedstock weight, tons including water

(2) $\text{C:N Ratio} = \dfrac{\text{Wt. of "C" in Material A + "C" in B + "C" in Material C}}{\text{Wt. of "N" in Material A + "N" in B + "N" in C}}$

Required amount of material "A" per pound of material "B" based on the desired moisture content.

$$\text{Weight of material A} = \frac{\text{Moisture of material B - Desired moisture}}{\text{Desired moisture - Moisture of material A}}$$

Now check the C:N ratio using equation (2).

Required amount of material "A" per pound of material "B" based on the desired C:N Ratio.

$$\text{Weight of Material A} = \frac{\% N^b}{\% N^a} \times \frac{(R - R^b)}{(R^a - R)} \times \frac{(1 - M^b)}{(1 - M^a)}$$

Symbols:

A = pounds of material A per pound of material B
M^a = Moisture content of material A
M^b = Moisture content of material B
% N^a = % Nitrogen in material A
% N^b = % Nitrogen in material B
R = Desired C:N ratio
R^a = C:N ratio of material A
R^b = C:N ratio of material B

Now check the moisture content using equation (1)

Additional Resources

Internet sites represent a vast resource of information. Using one of the search engines on the Internet such as Google or DuckDuckGo, find more information by searching for words, phrases, or websites: Always use caution when searching for information on the Internet. Follow guidelines to ensure the accuracy and reliability of the information you find.

Search words: Mesophilic, Thermophilic, aerobic composting,

Self-Assessment

1. List the three primary sources of odors at a composting facility.

2. Name the four primary nutrients required by microorganisms involved in composting.

3. Describe how the pH level affects composting.

4. What is the efficient moisture level for compost and why?

Notes

28 Landfill Gas Energy Basics

Major Concept

Landfill gas (LFG) is a natural byproduct of the decomposition of organic material under anerobic conditions. LFG contains roughly 50 to 55% methane and 45 to 50% carbon dioxide (CO_2), with less than 1% non-methane organic compounds.

Objectives

- Describe the benefits of harnessing landfill gas
- Define landfill gas and how it is created
- List the uses of landfill gas
- Identify the economic benefits of LFG

Key Terms

- Anaerobic
- Flare
- Greenhouse gases
- Landfill cell
- Leachate
- Municipal solid waste

Chapter Resource

Complementary *full color* illustrations, photos, charts, and graphs are available for Chapter 28 by following this URL: https://tinyurl.com/7vxzdaks This digital resource will enhance your understanding of the chapter concepts.

Introduction

- Harnessing the power of Landfill Gas (LFG) energy provides environmental and economic benefits to landfills, energy users and the community.

   o Working together, landfill owners, energy service providers, businesses, state agencies, local governments, communities, and other stakeholders can develop successful LFG energy projects that:

      ✓ Reduce emissions of greenhouse gases (GHGs) that contribute to global climate change
      ✓ Offset the use of non-renewable resources
      ✓ Help improve local air quality
      ✓ Provide revenue for landfills
      ✓ Reduce energy costs for users of LFG energy
      ✓ Create jobs and promote investment in local businesses

- The selection describes the source and characteristics of LFG and presents basic information about the collection, treatment and use of LFG in energy recovery systems.

- This selection also includes a discussion of the status of LFG energy in the United States, a review of the benefits of LFG energy projects and a summary of the current federal regulatory framework.

- Finally, general steps to LFG energy project development are introduced.

What is LFG?

- LFG is a natural byproduct of the decomposition of organic material in anerobic (without oxygen) conditions.

- LFG contains roughly 50 to 55% methane and 45 to 50% carbon dioxide (CO_2), with less than 1% non-methane organic compounds (NMOCs) and trace amounts of inorganic compounds.

    - Methane is a potent GHG 28 to 36 times more effective than carbon dioxide at trapping heat in the atmosphere over a 100-year period.

- LMOP uses a methane global warming potential (GWP) of 25 in program calculations to be consistent with and comparable to key Agency emission quantification programs such as the U.S. GHG Inventory.

- When municipal solid waste (MSW) is first deposited in a landfill, it undergoes an aerobic (with oxygen) decomposition stage when little methane is generated.

    - Then, typically within less than one-year, anaerobic conditions are established and methane-producing bacteria begin to decompose the waste and generate methane.

    - Approximately 300 million tons of MSW are generated in the United States with less than 49% of that deposited in landfills.

- One million tons of MSW produces roughly 300 cubic feet per minute (cfm) of LFG and continues to produce LFG for as many as 20 to 30 years after it has been landfilled.

- With a heating value of about 500 British thermal units (Btu) per standard cubic foot, LFG is a good source of useful energy, normally through the operation of engines or turbines.

- Many landfills collect and use LFG voluntarily to take advantage of this renewable energy resource while also reducing GHG emissions.

Modern landfill

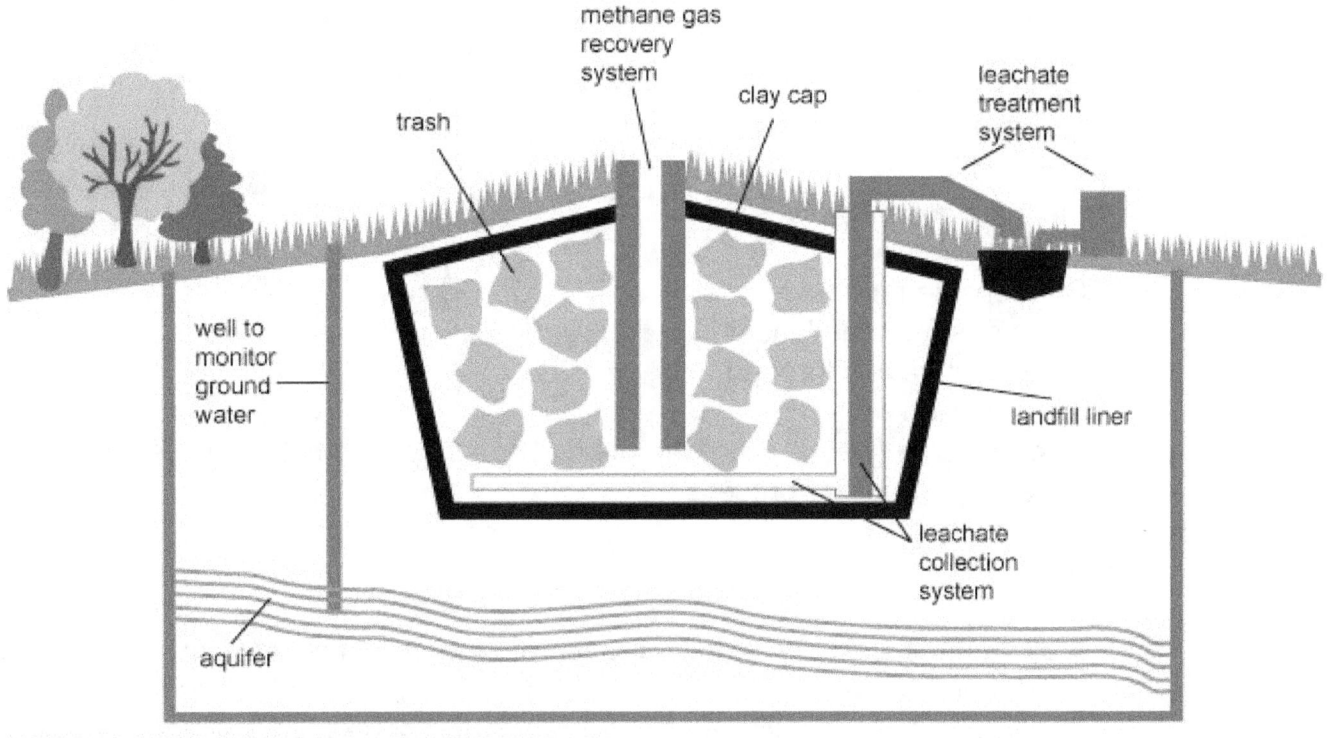

Source: Adapted from National Energy Education Development Project (public domain)

LFG Collection and Flaring

- LFG collection typically begins after a portion of the landfill (known as a "cell") is closed to additional waste placement.

  o Collection systems can be configured as either vertical wells or horizontal trenches.
- Most landfills with energy recovery systems include a flare for the combustion of excess gas and for use during equipment downtimes.

- Each of these components is described below, followed by a brief discussion of collection system and flare costs.

  o Gas Collection Wells and Horizontal Trenches

    ✓ The most common method of LFG collection involves drilling vertical wells in the waste and connecting those wellheads to lateral piping that transports the gas to a collection header using a blower or vacuum induction system.
    ✓ Another type of LFG collection system uses horizontal piping laid in trenches in the waste. Horizontal trench systems are useful in deeper landfills and in areas of active filling.
    ✓ Some collection systems involve a combination of vertical wells and horizontal collectors.

- ✓ Well-designed systems of either type are effective in collecting LFG.
- ✓ The design chosen depends on site-specific conditions and the timing of LFG collection system installation.

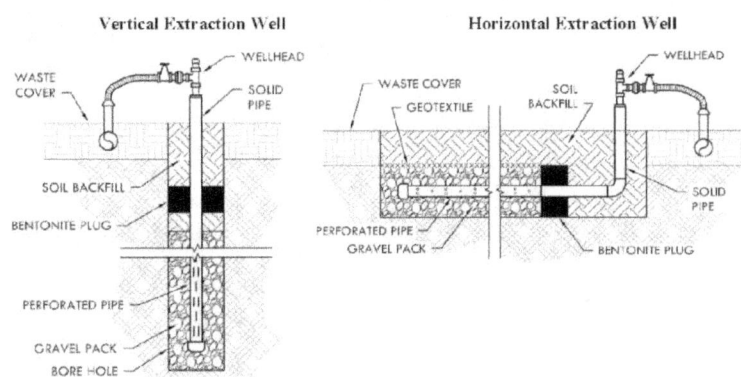

o Condensate collection

- ✓ Condensate forms when warm gas from the landfill cools as it travels through the collection system.
- ✓ If condensate (water) is not removed, it can block the collection system and disrupt the energy recovery process.

o Blower

- ✓ A blower is necessary to pull the gas from the collection wells into the collection header and convey the gas to downstream treatment and energy recovery systems.
- ✓ The size, type, and number of blowers needed depend on the gas flow rate and distance to downstream processes.

o Flare

- ✓ A flare is a device for igniting and burning the LFG.
- ✓ Flares are a component of each energy recovery option because they may be needed to control LFG emissions during startup and downtime of the energy recovery system and to control gas that exceeds the capacity of the energy conversion equipment.
- ✓ In addition, a flare is a cost-effective way to gradually increase the size of the energy generation system at an active landfill.
- ✓ As more waste is placed in the landfill and the gas collection system is expanded, the flare is used to control excess gas between energy conversion system upgrades (for example, before the addition of another engine) to prevent methane from being released into the atmosphere.
- ✓ Flare designs include open (or candlestick) flares and enclosed flares.

- ✓ Enclosed flares are more expensive but may be preferable (or required by state regulations) because they provide greater control of combustion conditions, allow for stack testing, and might achieve slightly higher combustion efficiencies (higher methane destruction rates) than open flares.
  - ➢ They can also reduce noise and light nuisances.

LFG Treatment

- Using LFG in an energy recovery system usually requires some treatment of the LFG to remove excess moisture, particulates, and other impurities.

- The type and extent of treatment depend on site-specific LFG characteristics and the type of energy recovery system employed.

- Boilers and most internal combustion engines generally require minimal treatment (usually dehumidification, particulate filtration, and compression).

  - Some internal combustion engines and many gas turbine and microturbine applications also require siloxane and hydrogen sulfide removal using adsorption beds, biological scrubbers, and other available technologies after the dehumidification step.

- An LFG energy project, including LFG collection, is a fairly extensive treatment system with the energy recovery system generating both electricity and heat.

  - Most LFG energy projects produce either electricity or heat, although a growing number of combined head and power (CHP) systems produce both.

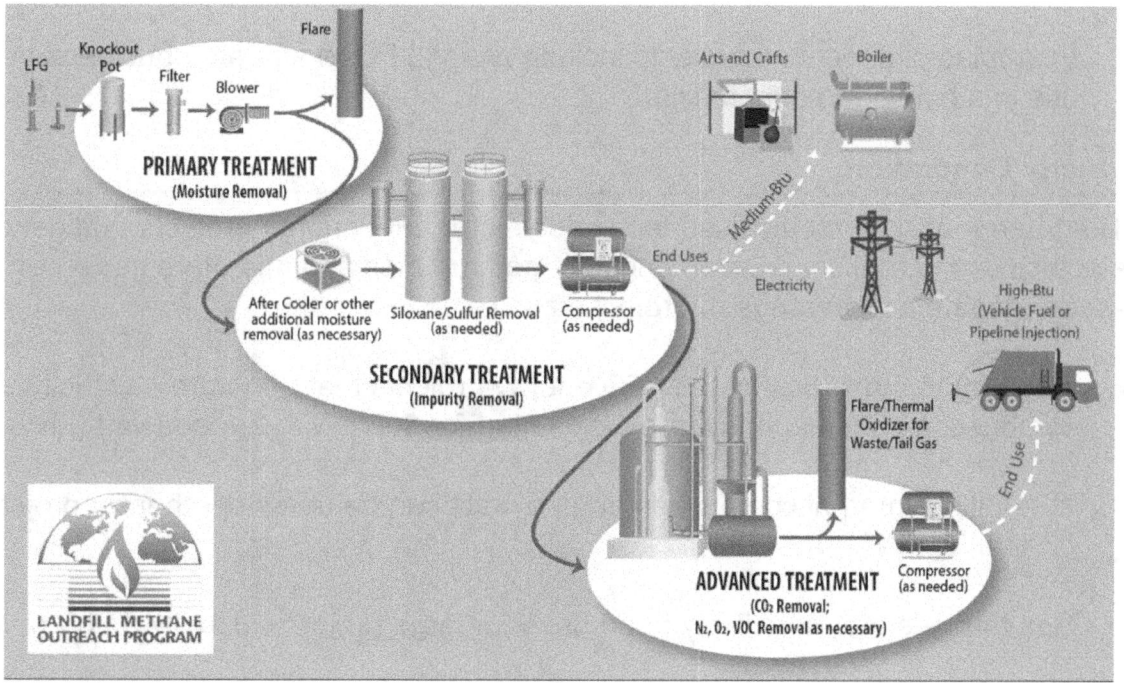

- The cost of gas treatment depends on the gas purity requirements of the end use application.

    o The cost of a system to filter the gas and remove condensate for direct use of medium-BTU gas or for electric power production is considerably less than the cost of a system that must also remove contaminants such as siloxane and sulfur that are present at elevated levels in some LFG.

Uses of LFG

- LFG energy projects first came on the scene in the mid-to late-1970s and increased notably during the 1990s as a track record for efficiency, dependability, and cost savings was demonstrated.

- The enactment of federal tax credits and regulatory requirements for LFG collection and control for larger landfills also helped to spur the growth of LFG energy projects, as did other factors such as increased concerns about how methane emissions contribute to global climate change and market demands for renewable energy options.

- LMOP's Landfill and LFG Energy Project Database, which tracks the development of U.S. LFG energy projects and landfills with project development potential, indicates that, as of March 2021, there are 550 operational LFG energy projects in the United States and 481 landfills that are good candidates for projects.

    o Roughly three-quarters of these projects generate electricity, while the remainder are either direct-use projects where the LFG is used for its thermal capacity or upgraded LFG projects where the LFG is cleaned to a level similar to natural gas.

    o Examples of direct-use projects include piping LFG to a nearby business or industry for use in a boiler, furnace, or kiln.

Electricity Generation

- The three most commonly used technologies for LFG energy projects that generate electricity – internal combustion engines, gas turbines and microturbines – can accommodate a wide range of project sizes.

    o Most (more than 75%) of the LFG energy projects that generate electricity use internal combustion engines, which are well-suited for 800 kW to 3-megawatt (MW) projects.

        ✓ Multiple internal combustion engine units can be used together for projects larger than 3 MW.

    o Gas turbines are more likely to be used for large projects, usually 5 MW or larger.

- - Microturbines, as their name suggests, are much smaller than gas turbines, with a single unit having between 30 and 250 kW in capacity, and are generally used for projects smaller than 1 MW.

    - ✓ Small internal combustion engines are also available for projects in this size range.

- CHP applications, also known as cogeneration projects, provide greater overall energy efficiency and are growing in number.

  - In addition to producing electricity, these projects recover and beneficially use the heat from the unit combusting the LFG.
  -

- LFG energy CHP projects can use internal combustion engines, gas turbines or microturbine technologies.

- Other LFG electricity generation technologies include boiler/steam turbines and combine cycle applications.

  - In boiler/steam turbine applications, LFG is combusted in a large boiler to generate steam that powers a turbine to create electricity.

- Combined cycle applications combine a gas turbine with a steam turbine, so that the gas turbine combusts the LFG and the steam turbine uses the steam generated from the gas turbine's exhaust to create electricity.

- Boiler/steam turbine and combined cycle applications tend to be larger in scale than the majority of LFG electricity projects that use internal combustion engines.

Direct Use

- Direct use of LFG can offer a cost-effective alternative for fueling combustion or heating equipment at facilities located within approximately 5 miles of a landfill.

- In some situations, longer pipelines may be economically feasible based on the amount of LFG collected, the fuel demand of the end user and the price of the fuel the LFG will replace.

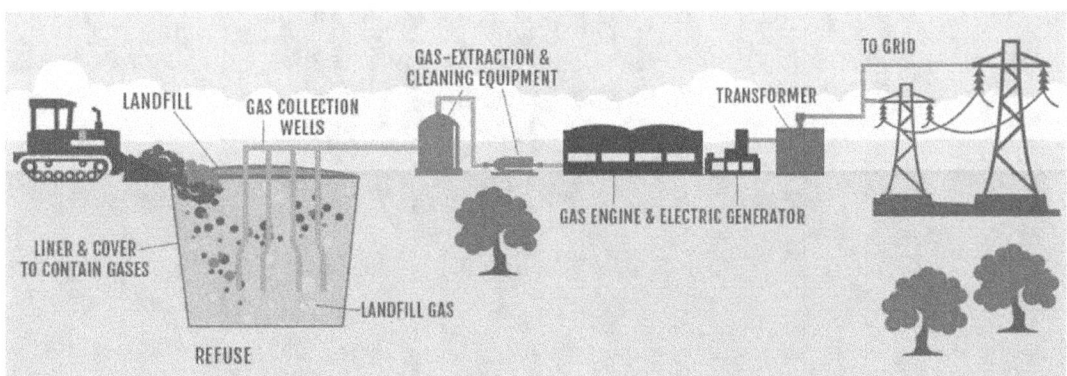

- Some manufacturing plants have chosen to locate near a landfill for the express purpose of using LFG as a renewable fuel that is cost-effective when compared with natural gas.

- The number and diversity of direct-use LFG applications is continuing to grow. Project types include:

  - **Boilers**, which are the most common type of direct use and can often be easily converted to the LFG alone or in combination with fossil fuels.

  - **Direct thermal applications**, which include kilns (cement, pottery, or brick), sludge dryers, infrared heaters, paint shop oven burners, tunnel furnaces, process heaters and blacksmithing forges, to name a few. LFG has also found a home in a few greenhouse operations.

  - **Leachate evaporation**, in which a combustion device that uses LFG is used to evaporate leachate (the liquid that percolates through a landfill). Leachate evaporation can reduce the cost of treating and disposing of leachate.

- The creation of pipeline-quality, of high-BTU, gas from LFG is becoming more prevalent.

  - In this process, LFG is cleaned and purified (carbon dioxide and impurities removed) until it is at the quality that can be directly injected into a natural gas pipeline.

- Also growing in popularity are projects in which LFG provides heat for processes that create alternative fuels (such as biodiesel or ethanol).

  - In some cases, LFG is directly used as feedstock for an alternative fuel (for example, compressed natural gas [CNG], liquefied natural gas [LNG], or methanol).

  - Only a handful of these projects are currently operational, but several more are in the construction or planning stages.

Operational LFG Energy Projects by Type (August 2020)
SOURCE: US EPA LMOP

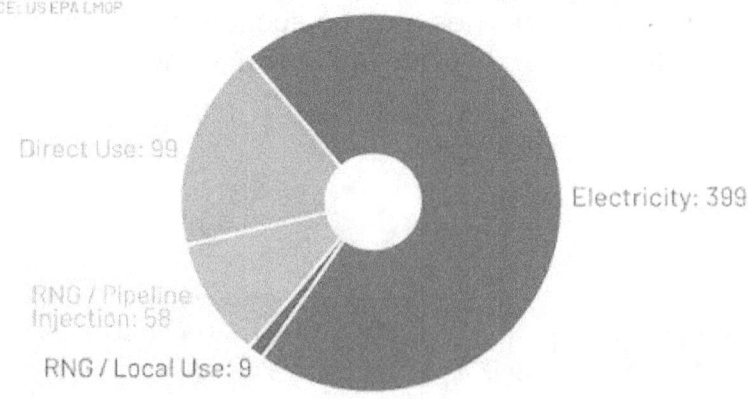

Direct Use: 99
Electricity: 399
RNG / Pipeline Injection: 58
RNG / Local Use: 9

Environmental and Economic Benefits of LFG Energy Recovery

- Developing LFG energy projects is an effective way to reduce GHG emissions, improve local air quality and control odors.

- This section highlights the numerous environmental and economic benefits that LFG energy projects provide to the community, the landfill, and the energy end user.

- Environmental Benefits

    o MSW landfills are the third-largest human-caused source of methane emissions in the United States.

    o Methane is a potent heat-trapping gas (25 times stronger than carbon dioxide over a 100-year period) and has a short atmospheric life (~12 years).

        ✓ Because methane is both potent and short-lived, reducing methane emissions from MSW landfills is one of the best ways to lessen the human impact on global climate change.
        ✓ In addition, all landfills generate methane, so there are many opportunities to reduce methane emissions by flaring or collecting LFG for energy generation.

- Direct GHG Reductions

    o During its operational lifetime, an LFG energy project will capture an estimated 60 to 90% of the methane created by a landfill, depending on system design and effectiveness.

    o The methane captured is converted to water and carbon dioxide when the gas is burned to produce electricity or heat.

- Indirect GHG Reductions

    o Producing energy from LFG displaces the use of non-renewable resources (such as coal, oil, or natural gas) that would be needed to produce the same amount of energy.

    o This displacement avoids GHG emissions from fossil fuel combustion by an end user facility or power plant.

- Direct and Indirect Reduction of Other Air Pollutants

    o The capture and use of LFG at a landfill improves local air quality in many ways. For example:

        ✓ NMOCs that are present at low concentrations in LFG are destroyed or converted during combustion, which reduces possible health risks.

- ✓ For electricity projects, the avoidance of fossil fuel combustion at utility power plants means that fewer pollutants are released into the air, including sulfur dioxide (which is a major contributor to acid rain), particulate matter (a respiratory health concern), nitrogen oxides (which can contribute to local ozone and smog formation) and trace hazardous air pollutants.
- ✓ LFG energy use helps to avoid the use of limited, non-renewable resources such as coal and oil.
- ✓ Although the equipment that burns LFG to generate electricity generates some emissions, including nitrogen oxides, the overall environmental benefits achieved from LFG energy projects are significant because of the direct methane reductions, the indirect carbon dioxide reductions, and the direct and indirect reduction in other air pollutant emissions.

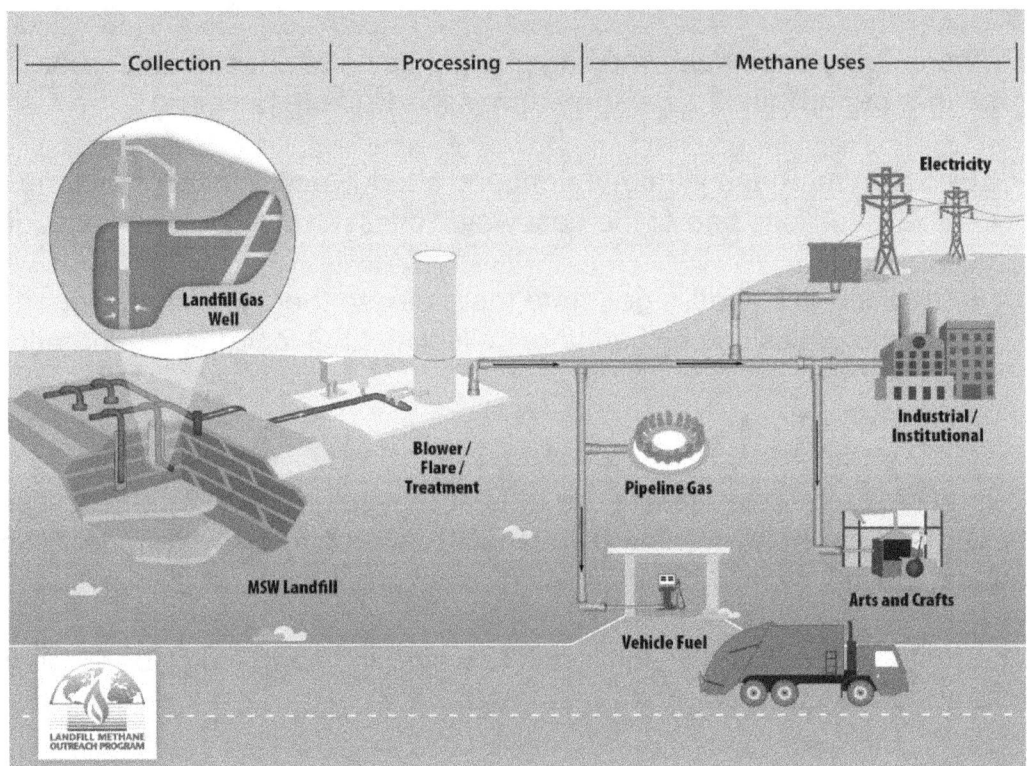

- Other Environmental Benefits

  o Collecting and combusting LFG improves the quality of the surrounding community by reducing landfill odors that are usually caused by sulfates in the gas.

  o Collecting LFG also improves safety by reducing gas migration to structures, where trapped or accumulated gas can create explosion hazards.

- Economic Benefits

  o For the Landfill Owner

    ✓ Landfill owners can receive revenue from the sale of LFG to a direct end user or pipeline, or from the sale of electricity generated from LFG to the local power grid.
    ✓ Depending on who owns the rights to the LFG and other factors, a landfill owner may also be eligible for revenue from renewable energy certificates (RECs), tax credits and incentives, renewable energy bonds, and GHG emissions trading.
    ✓ All these potential revenue sources can help offset gas collection system and energy project costs for the landfill owner,
    ✓ For example, if the landfill owner is required to install a gas collection and control system, using the LFG as an energy resource can help pay down the capital cost required for the control system installation.

  o For the Community

    ✓ LFG energy project development can greatly benefit the local economy.
    ✓ Temporary jobs are created for the construction phase, while design and operation of the collection and energy generation systems create long-term jobs.
    ✓ LFG energy projects involve engineers, construction firms, equipment vendors, and utilities or end users of the power produced.
    ✓ Some materials for the overall project may be purchased locally, and often local firms are used for construction, well drilling, pipeline installation and other services.
    ✓ In addition, lodging and meals for the workers provide a boost to the local economy.
    ✓ Some of the money paid to workers and local businesses by the LFG energy project is spent within the local economy on goods and services, resulting in indirect economic benefits.
    ✓ In some cases, LFG energy projects have led new businesses (such as brick and ceramics plants, greenhouses, or craft studios) to locate near the landfill to use LFG.
    ✓ These new businesses add depth to the local economy.

Overview of the Regulatory Framework

- Landfills and LFG energy projects can be subject to air quality, solid waste and water quality regulations and permitting requirements.

- State and local governments typically develop their own regulations for carrying out the federal mandates; therefore, specific requirements differ among states.

- In addition, project developers should contact relevant federal agencies and state agencies for more detailed, current information and to obtain applications for various types of construction and operating permits.
- An overview of the federal regulatory framework is presented below.

- It is important for project developers to review applicable requirements and regulations. Project developers are responsible for ensuring compliance with applicable regulations.

Steps to Developing LFG Energy Projects

- The following section provides a basic overview of nine general steps involved in developing an LFG energy project.

- Step 1: Estimate LFG Recovery Potential and Perform Initial Assessment

    o The first step is to determine whether the landfill is likely to produce enough methane to support an energy recovery project. Initial screening criteria include:

       ✓ Does the landfill contain at least 1 million tons of MSW?
       ✓ Does the landfill have a depth of 50 feet or more?
       ✓ Is the landfill open or recently closed?
       ✓ Does the site receive at least 25 inches of precipitation annually?
       ✓ Does the landfill contain enough organic content to generate sufficient LFG?

    o Landfills that meet these criteria are likely to generate enough gas to support an LFG energy project.

    o It is important to note that these are only ideal conditions, many successful LFG energy projects have been developed at smaller, older, or more arid landfills.

    o If it is determined that the energy recovery option is viable, then it is important to estimate the amount of recoverable gas that will be available over time.

       ✓ EPA's LandGEM can provide a more detailed analysis of LFG generation potential.

    o An important factor for LFG generation is the organic content of the MSW.

    o Waste composed of higher organic content will produce more LFG than waste with lower organic content.

    o Construction and demolition (C&D) landfills, for example, are not expected to generate large quantities of LFG and are often not viable for an energy generation system.

- Step 2: Evaluate Project Economics

    o The next step is to perform a detailed economic assessment of converting LFG into a marketable energy product such as electricity, steam, boiler fuel, vehicle fuel, or pipeline-quality gas.

    o A variety of technologies can be sued to maximize the value of LFG.

- The best configuration for a particular landfill will depend on a number of factors, including the existence of an available energy market, project costs, potential revenue sources and other technical considerations.

- LMO's LFGcost-Web tool can help with preliminary economic evaluation.

- Step 3: Establish Project Structure

    - Implementation of a successful LFG energy project begins with identifying the appropriate management structure.

    - For example, options for managing an LFG energy project include:

        - ✓ The landfill owner develops and manages the project internally.
        - ✓ The landfill owner teams with an external project developer so that the developer finances, constructs, owns and operates the project.
        - ✓ The landfill owner teams with partners (such as an equipment supplier or energy end user).

    - LMOP can assist with project partnering by identifying potential matches and distributing RFPs.

    - The LMOP Locator tool available for download online allows users to search for facilities that could potentially benefit from LFG or search for landfills that could potentially provide LFG to an interested party.

- Step 4: Draft Development Contract

    - The terms of LFG energy project partnerships should be formalized in a development contract.

    - The contract identifies which partner owns the gas rights and the rights to potential emission reductions.

    - The contract also establishes each partner's responsibilities, including design, installation and operation and maintenance.

    - Contracting with a developer is a complex issue, and each contract will be different depending on the specific nature of the project and the objectives and limitation of the participants.

- Step 5: Negotiate Energy Sales Contract (Off-Take Agreement)

    - The LFG energy project owner and the end user negotiate an energy sales contract that specifies the amount of gas or power to be delivered by the project owner to the end user and the price to be paid by the end user for the gas or power.

- The terms of the energy sales contract typically dictate the success of failure of the LFG energy project because they secure the project's source of revenue.

- Therefore, successfully obtaining this contract is a crucial milestone in the project development process.

- Negotiating an energy sales contract involves the following actions:
    - ✓ Evaluating the end user's need for gas or power
    - ✓ Preparing a draft offer contract
    - ✓ Developing the project design and pricing
    - ✓ Preparing and presenting a bid package
    - ✓ Reviewing contract terms and conditions
    - ✓ Signing the contract

- Because contract negotiation is often a complex process, owners and developers should consult an expert for further information and guidance.

- Step 6: Secure Permits and Approvals

    - Obtaining the required permits (environmental, siting and others) is an essential step in the development process.

    - Permit conditions often affect project design, and neither construction nor operation may begin until the appropriate permits are in place.

    - The process of permitting and LFG energy project can take anywhere from 6 to 18 months (or longer) to complete, depending on the location and recovery technology.

    - LFG energy projects must comply with federal regulations related to both the control of LFG emissions and the control of air emissions from the energy conversion equipment.

    - The landfill owner should contact and meet with regulatory authorities to identify requirements and educate the local officials, landfill neighbors, and nonprofit and other public interest and community groups about the benefits of the project.

    - LMOP's State Agencies page lists websites for state organizations that can provide useful information regarding state-specific regulations and permits.

- Step 7: Assess Financing Options

    - Financing an LFG energy project is one of the most important and challenging tasks facing a landfill owner or project developer.

- A number of potential financing sources are available, including equity investors, loans from investment companies or banks and municipal bonds.

- Five general categories of financing methods may be available to LFG energy projects:
  - ✓ Private equity financing
  - ✓ Project financing
  - ✓ Municipal bond funding
  - ✓ Direct municipal financing
  - ✓ Lease financing

- In addition to financing options, there are a variety of financial incentives available at the federal and state levels.

- **Step 8: Contract for Engineering, Procurement, and Construction (EPC) and O&M Services**

  - The construction and operation of LFG energy projects is complex, so it may be in the interest of the landfill owner to hire a firm with proven experience gained over the course of implementing similar projects.

  - Landfill owners who choose to contract with EPC and O&M firms should solicit bids from several EPC or O&M contractors before a contract is negotiated.

  - In most cases, the selected EPC or O&M contractor conducts the engineering design, site preparation and plant construction, and startup testing for the LFG energy project.

- **Step 9: Install Project and Start Up**

  - The final phase of implementation is the start of commercial operations.

  - This phase is often commemorated with ribbon-cutting ceremonies, public tours and press releases.

Additional Resources

Internet sites represent a vast resource of information. Usingone of the search engines on the Internet such as Google or DuckDuckGo, find more information by searching for words, phrases, or websites: Always use caution when searching for information on the Internet. Follow guidelines to ensure the accuracy and reliability of the information you find.

Search words: gas collection wells, flaring, condensate collection, direct thermal applications, leachate evaporation, Milner Butte Landfill

Website: https://www.epa.gov/lmop/basic-information-about-landfill-gas

Self-Assessment

1. Name the three most commonly used technologies for LFG energy products.

2. List the three types of projects that are direct use LGF applications.

3. Describe how LFG energy can help a community.

4. Identify an LFG project in your area and write a summary.

29 Recycling

Major Concept

The goals of recycling are to change waste materials into new products to prevent waste of potentially useful materials, to reduce the consumption of fresh raw materials, to reduce energy usage, to reduce air pollution (from incineration) and water pollution (from landfilling) by reducing the need for "conventional" waste disposal, and to lower greenhouse gas emissions.

Objectives

- Describe the process of recycling
- Trace the history of recycling
- List the ways recycled items are collected
- Identify the seven types of recycling codes for plastics
- Explain the environmental impact of recycling
- Evaluate the cost-benefit of recycling

Key Terms

- Emergy
- E-waste
- Exergy
- Recyclates
- Recyclebots

Chapter Resource

Complementary *full color* illustrations, photos, charts, and graphs are available for Chapter 29 by following this URL: https://tinyurl.com/7vxzdaks This digital resource will enhance your understanding of the chapter concepts.

Introduction

- Recycling is a process to change waste materials into new products to prevent waste of potentially useful materials, reduce the consumption of fresh raw materials, reduce energy usage, reduce air pollution (from incineration) and water pollution (from landfilling) by reducing the need for "conventional" waste disposal, and lower greenhouse gas emissions.

- Recycling is a key component of modern waste reduction and is the third component of the "Reduce, Reuse and Recycle" waste hierarchy.
  - There are some ISO standards related to recycling such as ISO 15270:2008 for plastics waste and ISO 14001:2004 for environmental management control of recycling practice.

- Recyclable materials include many kinds of glass, paper, metal, plastic, textiles, and electronics.

- The composting or other reuse of biodegradable waste – such as food or garden waste – is also considered recycling.

- Materials to be recycled are either brought to a collection center or picked up from the curbside, then sorted, cleaned, and reprocessed into new materials bound for manufacturing.

    - In the strictest sense, recycling of a material would produce a fresh supply of the same material – for example, used office paper would be converted into new office paper, or used foamed polystyrene into new polystyrene.

    - However, this is often difficult or too expensive (compared with producing the same product from raw materials or other sources), so "recycling" of many products or materials involves their reuse in producing different materials (e.g., paperboard) instead.

- Another form of recycling is the salvage of certain materials from complex products, either due to their intrinsic value (e.g., lead from car batteries, or gold from computer components), or due to their hazardous nature (e.g., removal and reuse of mercury from various items).

- Critics dispute the net economic and environmental benefits of recycling over its costs and suggest that proponents of recycling often make matters worse and suffer from confirmation bias.

    - Specifically, critics argue that the costs and energy used in collection and transportation detract from (and outweigh) the costs and energy saved in the production process; also, that the jobs produced by the recycling industry can be a poor trade for the jobs lost in logging, mining, and other industries associated with virgin production; and materials such as paper pulp can only be recycled a few times before material degradation prevents further recycling.

    - Proponents of recycling dispute each of these claims, and the validity of arguments from both sides has led to enduring controversy.

History

- Recycling has been a common practice for most of human history, with recorded advocates as far back as Plato in 400 BC.

    - During periods when resources were scarce, archaeological studies of ancient waste dumps show less household waste (such as ash, broken tools, and pottery) – implying more waste was being recycled in the absence of new material.

    - In pre-industrial times, there is evidence of scrap bronze and other metals being collected in Europe and melted down for perpetual reuse.

    - In Britain, dust and ash from wood and coal fires was collected by 'dustmen' and down cycled as a base material used in brick making.

- The main driver for these types of recycling was the economic advantage of obtaining recycled feedstock instead of acquiring virgin material, as well as a lack of public waste removal in ever more densely populated areas.

- In 1813, Benjamin Law developed the process of turning rags into 'shoddy' and 'mungo' wool in Batley, Yorkshire.

    - This material combined recycled fibers with virgin wool.

- The West Yorkshire shoddy industry in towns such as Batley and Dewsbury, lasted from the early 19th century to at least 1914.

- Industrialization spurred demand for affordable materials; aside from rags, ferrous scrap metals were coveted as they were cheaper to acquire than was virgin ore.

- Railroads both purchased and sold scrap metal in the 19th century, and the growing steel and automobile industries purchased scrap in the early 20th century.

    - Many secondary goods were collected, processed, and sold by peddlers who combed dumps, city streets, and went door to door looking for discarded machinery, pots, pans, and other sources of metal.

    - By World War I, thousands of such peddlers roamed the streets of American cities, taking advantage of market forces to recycle post-consumer materials back into industrial production.

- Beverage bottles were recycled with a refundable deposit at some drink manufacturers in Great Britain and Ireland around 1800, notably Schweppes.

    - An official recycle system with refundable deposits was established in Sweden for bottles in 1884 and aluminum beverage cans in 1982, by law, leading to a recycling rate for beverage containers of 84-99% depending on type, and average use of a glass bottle is over 20 refills.

World War II and Post-War

- Wartime recycling was a highlight throughout World War II.
- During the war, financial constraints, and significant material shortages due to war efforts made it necessary for countries to reuse goods and recycle materials.

- These resource shortages caused by the world wars, and other such world-changing occurrences, greatly encouraged recycling.

- The struggles of war claimed much of the material resources available, leaving little for the civilian population.

- It became necessary for most homes to recycle their waste, as recycling offered an extra source of materials allowing people to make the most of what was available to them.

- Recycling household materials meant more resources for war efforts and a better chance of victory.

- Massive government promotion campaigns were carried out in the home front during World War II in every country involved in the war, urging citizens to donate metals and conserve fiber, as a matter of patriotism.

- A considerable investment in recycling occurred in the 1970s, due to rising energy costs.

  - Recycling aluminum uses only 5% of the energy required by virgin production; glass, paper, and metals have less dramatic but significant energy savings when recycled feedstock is used.

- As of 2014, the European Union has about 50% of world share of the waste and recycling industries, with over 60,000 companies employing 500,000 persons, with a turnover of €24 billion.

  - Countries have to reach recycling rates of at least 50%, while the lead countries are around 65% and the EU average is 39% as of 2013.

Legislation

- For a recycling program to work, having a large and stable supply of recyclable material is crucial.

- Three legislative options have been used to create such a supply: mandatory recycling collection, container deposit legislation, and refuse bans.

  - Mandatory collection laws set recycling targets for cities to aim for, usually in the form that a certain percentage of a material must be diverted from the city's waste stream by a target date.

    ✓ The city is then responsible for working to meet this target.

  - Container deposit legislation involves offering a refund for the return of certain containers, typically glass, plastic, and metal.

- - - ✓ When a product in such a container is purchased, a small surcharge is added to the price.
    - ✓ This surcharge can be reclaimed by the consumer if the container is returned to a collection point.
    - ✓ These programs have been successful, often resulting in an 80% recycling rate.
    - ✓ Despite such good results, the shift in collection costs from local government to industry and consumers has created strong opposition to the creation of such programs in some areas.

  - A third method of increase supply of recyclates is to ban the disposal of certain materials as waste, often including used oil, old batteries, tires, and garden waste.

    - ✓ One aim of this method is to create a viable economy for proper disposal of banned products.
    - ✓ Care must be taken that enough of these recycling services exist, or such bans simply lead to increased illegal dumping.

- Government-mandated demand legislation has also been used to increase and maintain a demand for recycled materials.

- Four methods of such legislation exist: minimum recycled content mandates, utilization rates, procurement policies, and recycled product labeling.

  - Both minimum recycled content mandates and utilization rates increase demand directly by forcing manufacturers to include recycling in their operations.

  - Content mandates specify that a certain percentage of a new product must consist of recycled material.

  - Utilization rates are a more flexible option: industries are permitted to meet the recycling targets at any point of their operation or even contract recycling out in exchange for [trade]able credits.

  - Opponents to both of these methods point to the large increase in reporting requirements they impose, and claim that they rob industry of necessary flexibility.

- Governments have used their own purchasing power to increase recycling demand through what are called "procurement policies."

  - These policies are either "setasides," which earmark a certain amount of spending items are purchased.

- - Additional regulations can target specific cases: in the United States, for example, the Environmental Protection Agency mandates the purchase of oil, paper, tires, and building insulation from recycled or re-refined sources whenever possible.

  - Guidelines for reduce, reuse, recycle from the EPA: https://www.epa.gov/recycle.

- The final government regulation towards increased demand is recycled product labeling.

  - When producers are required to label their packaging with amount of recycled material in the product (including the packaging), consumers are better able to make educated choices.

  - Consumers with sufficient buying power can then choose more environmentally conscious options, prompt producers to increase the amount of recycled material in their products, and indirectly increase demand.

  - Standardized recycling labeling can also have a positive effect on supply of recyclates if the labeling includes information on how and where the product can be recycled.

Recyclates

- Recyclate is a raw material that is sent to, and processed, in a waste recycling plant or materials recovery facility which will be used to form new products.

- The material is collected in various methods and delivered to a facility where it undergoes re-manufacturing so that it can be used in the production of new materials or products.

  - For example, plastic bottles that are collected can be re-used and made into plastic pellets, a new product.

- The quality of recyclates is recognized as one of the principal challenges that needs to be addressed for the success of a long-term vision of a green economy and achieving zero waste.

- Recyclate quality is generally referring to how much of the raw material is made up of target material compared to the amount of non-target material and other non-recyclable material.

  - Only target material is likely to be recycled, so a higher amount of non-target and non-recyclable material will reduce the quantity of recycling product.

  - A high proportion of non-target and non-recyclable material can make it more difficult for re-processors to achieve 'high-quality' recycling.

- - If the recyclate is of poor quality, it is more likely to end up being downcycled or, in more extreme cases, sent to other recovery options or landfill.
    - ✓ For example, to facilitate the re-manufacturing of clear glass products there are tight restrictions for colored glass going into the remelt process.
- The quality of recyclate not only supports high quality recycling, but it can also deliver significant environmental benefits by reducing, reusing, and keeping products out of landfills.
- High quality recycling can help support growth in the economy by maximizing the economic value of the waste material collected.
  - Higher income levels from the sale of quality recyclates can return value which can be significant to local governments, households, and businesses.
- Pursuing high quality recycling can also provide consumer and business confidence in the waste and resource management sector and may encourage investment in that sector.
- There are many actions along the recycling supply chain that can influence and affect the material quality of recyclate.
  - It begins with the waste producers who place non-target and non-recyclable wastes in recycling collection.
  - This can affect the quality of final recyclate streams or require further efforts to discard those materials at later stages in the recycling process.
- The different collection systems can result in different levels of contamination.
  - Depending on which materials are collected together, extra effort is required to sort this material back into separate streams and can significantly reduce the quality of the final product.
- Transportation and the compaction of materials can make it more difficult to separate material back into separate waste streams.
- Sorting facilities are not 100% effective in separating materials, despite improvements in technology and quality recyclate which can see a loss in recyclate quality.
- The storage of materials outside where the product can become wet can cause problems for reprocessors.
  - Reprocessing facilities may require further sorting steps to further reduce the amount of non-target and non-recyclable material.
  - Each action along the recycling path plays a part in the quality of recyclate.

- A number of different systems have been implemented to collect recyclates from the general waste stream.

- These systems lie along the spectrum of trade-off between public convenience and government ease and expense.

- The three main categories of collection are "drop-off centers," "buy-back centers," and "curbside collection."

Curbside Collection

- Curbside collection encompasses many subtly different systems, which differ mostly on where in the process the recyclates are sorted and cleaned.

- The main categories are mixed waste collection, commingled recyclables, and source separation.

- A waste collection vehicle generally picks up the waste.

    o At one end of the spectrum is mixed waste collection, in which all recyclates are collected, mixed in with the rest of the waste, and the desired material is then sorted out and cleaned at a central sorting facility.

    ✓ This results in a large amount of recyclable waste, paper especially, being too soiled to reprocess, but has advantages as well: the city need not pay for a separate collection of recyclates and no public education is needed.
    ✓ Any changes to which materials are recyclable is easy to accommodate as all sorting happens in a central location.

    o In a commingled or single-stream system, all recyclables for collection are mixed but kept separate from other waste.

    ✓ This greatly reduces the need for post-collection cleaning but does require public education on what materials are recyclable.

    o Source separation is the other extreme, where each material is cleaned and sorted prior to collection.

    ✓ This method requires the least post-collection sporting and produces the purest recyclates, but incurs additional operating costs for collection of each separate material.
    ✓ An extensive public education program is also required, which must be successful if recyclate contamination is to be avoided.

- Source separation used to be the preferred method due to the high sorting costs incurred by commingled (mixed waste) collection.

- Advances in sorting technology (see sorting below), however, have lowered this overhead substantially – many areas which had developed source separation programs have since switched to co-mingled collection.

Buy-Back Centers

- Buy-back centers differ in that the cleaned recyclates are purchased, thus providing a clean incentive for use, and creating a stable supply.

- The post-processed material can then be sold on, hopefully creating a profit.

- Unfortunately, government subsidies are necessary to make buy-back centers a viable enterprise.

  - According to the United States' National Waste & Recycling Association, it costs an average US $50 to process a ton of material, which can only be resold for US $30.

Drop-Off Centers and Distributed Recycling

- Drop-off centers require the waste producer to carry the recyclates to a central location.

  - This can be either an installed or mobile collection station or the reprocessing plant itself.

- They are the easiest type of collection to establish, but suffer from low and unpredictable throughput.

- For some waste materials such as plastic, recent technical devices called recyclebots enable a form of distributed recycling.

- Preliminary life-cycle analysis (LCA) indicates that such distributed recycling of HDPE to make filament of 3-D printers in rural regions is energetically favorable to either using virgin resin or conventional recycling processes because of reductions in transportation energy.

Sorting

- Once commingled recyclates are collected and delivered to a central collection facility, the different types of materials must be sorted.

- This is done in a series of stages, many of which involve automated processes such that a truckload of material can be fully sorted in less than an hour.
- Some plants can now sort the materials automatically, known as single-stream recycling.

- In plants a variety of materials are sorted such as paper, different types of plastics, glass, metals, food scraps, and most types of batteries.

  o A 30% increase in recycling rates has been seen in the areas where these plants exist.

- Initially, the commingled recyclates are removed from the collection vehicle and placed on a conveyor belt spread out in a single layer.

- Large pieces of corrugated fiberboard and plastic bags are removed by hand at this stage, as they can cause later machinery to jam.

- Next, automated machinery separates the recyclates by weight, splitting lighter paper and plastic from heavier glass and metal.

  o Cardboard is removed from the mixed paper, and the most common types of plastic, PET (#1) and HDPE (#2), are collected.

  o This separation is usually done by hand, but has become automated in some sorting centers.

    ✓ A spectroscopic scanner is used to differentiate between different types of paper and plastic based on the absorbed wavelengths, and subsequently divert each material into the proper collection channel.

- Strong magnets are used to separate out ferrous metals, such as iron, steel, and tin-plated cans ("tin cans").

- Nonferrous metals are ejected by magnetic eddy currents in which a rotating magnetic field induces an electric current around the aluminum cans, which in turn creates a magnetic eddy current inside the cans.

  o This magnetic eddy current is repulsed by the large magnetic field, and the cans are ejected from the rest of the recyclate stream.

- Finally, glass is sorted on the basis of its color: brown, amber, green, or clear.

  o It may either be sorted by hand, or via an automated machine that uses colored filters to detect different colors.
  o Glass fragments smaller than 10 mm across cannot be sorted automatically, and are mixed together as 'glass fines'.

- This process of recycling as well as reusing the recycled material proves to be advantageous for many reasons as it reduces the amount of waste sent to landfills, conserves natural resources, saves energy, reduces greenhouse gas emissions, and helps create new jobs.

- Recycled materials can also be converted into new products that can be consumed again such as paper, plastic, and glass.

- The City and County of San Francisco's Department of the Environment offers one of the best recycling programs to support its city-wide goal of Zero Waste by 2020. Although they were unable to meet the 2020 goal, they still continue to move forward.

    - San Francisco's refuse hauler, Recology, operates an effective recyclable sorting facility in San Francisco, which helped San Francisco reach a record-breaking diversion rate of 81%.

- Food packaging should no longer contain any organic matter.

    - Organic matter found in these needs to be placed in a biodegradable waste bin or can be buried in your garden.

- Since no trace of biodegradable material is best kept in the packaging before placing it in a trash bag, some packaging also needs to be rinsed.

Recycling Industrial Waste

- Although many government programs are concentrated on recycling at home, a large portion of waste is generated by industry.

- The focus of many recycling programs done by industry is the cost-effectiveness of recycling.

- The ubiquitous nature of cardboard packaging makes cardboard a commonly recycled waste product by companies that deal heavily in packaged goods, like retail stores, warehouses, and distributors of goods.

- Other industries deal in niche or specialized products, depending on the nature of the waste materials that are present.

    - The glass, lumber, wood pulp, and paper manufacturers all deal directly in commonly recycled materials.

    - However, old rubber tires may be collected and recycled by independent tire dealers for a profit.

- Levels of metals recycling are generally low.

- In 2010, the International Resource Panel, hosted by the United Nations Environment Programme (UNEP) published reports on metal stocks that exist within society and their recycling rates.

    o The Panel reported that the increase in the use of metals during the 20th and into the 21st century has led to a substantial shift in metal stocks from below ground to use in applications within society ground.

        ✓ For example, the in-use stock of copper in the USA grew from 73 to 238 kg per capita between 1932 and 1999.

    o The report authors observed that, as metals are inherently recyclable, the metal stocks in society can serve as huge mines above ground (the term "urban mining" has been coined with this idea in mind).

    o However, they found that the recycling rates of many metals are very low.

    o The report warned that the recycling rates of some rate metals sued in applications such as mobile phones, battery packs of hybrid cars and fuel cells, are so low that unless future end-of-life recycling rates are dramatically stepped up these critical metals will become unavailable for use in modern technology.

- The military recycles some metals. The U.S. Navy's Ship Disposal Program uses ship breaking to reclaim the steel of old vessels.

    o Ships may also be sunk to create an artificial reef.
    o Uranium is a very dense metal that has qualities superior to lead and titanium for many military and industrial uses.
    o The uranium left over from processing it into nuclear weapons and fuel for nuclear reactors is called depleted uranium, and it is used by all branches of the U.S. military for use as armor-piercing shells and shielding.

- The construction industry may recycle concrete and old road surface pavement, selling their waste materials for profit.

- Some industries like the renewable energy industry and solar photovoltaic technology in particular, are being proactive in setting up recycling policies even before there is considerable volume to their waste streams, anticipating future demand during their rapid growth.

- Recycling of plastics is more difficult, as most programs can't reach the necessary level of quality.

- Recycling of PVC often results in down cycling of the material, which means only products of lower quality standard can be made with the recycled material.

    - A new approach which allows an equal level of quality is the Vinyloop process.

        - ✓ It was used after the London 2012 Olympics to fulfill the PVC Policy.

e-Waste Recycling

- E-waste is a growing problem, accounting for 20-50 million tons of microprocessors retrieved from the waste stream of global waste per year according to the EPA.

- Many recyclers do not recycle e-waste or do not do so responsibly.

- The e-Stewards certification was created to ensure recyclers are held to the highest standards for environmental responsibility and to help consumers identify responsible recyclers.

    - e-Cycle, LLC, was the first mobile recycling company to be e-Stewards certified.

- In such a shop, TVs, monitors, mobile phones, and computers are typically reused or repaired.

    - If broken, they may be disassembled for parts still having high value.

    - Other e-waste is shredded to ~100 mm pieces and manually checked to separate out toxic batteries and capacitors which contain poisonous metals.

    - The remaining pieces are further shredded to ~10 mm and passed under a magnet to remove ferrous metals.

    - Next an eddy current ejects non-ferrous metals, which are sorted by density either by centrifuge or vibrating plates.

        - ✓ Precious metals can be dissolved in acid, sorted, and smelted into ingots.
        - ✓ The remaining glass and plastic fractions are separated by density and sold to re-processors.

- TVs and monitors must be manually disassembled to removed either toxic lead in CRTs or the mercury in flat screens.

Plastic Recycling

- Plastic recycling is the process of recovering scrap or waste plastic and reprocessing the material into useful products, sometimes completely different in form from their original state.

    - For instance, this could mean melting down soft drink bottles and then casting them as plastic chairs and tables.

- Physical Recycling

    - Some plastics are remelted to form new plastic objects.

        - For example, PET water bottles can be converted into clothing grade polyester.

    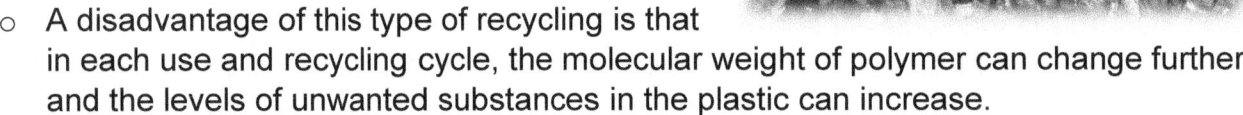

    - A disadvantage of this type of recycling is that in each use and recycling cycle, the molecular weight of polymer can change further and the levels of unwanted substances in the plastic can increase.

- Chemical Recycling

    - For some polymers it is possible to convert them back into monomers.

        - For example, PET can be treated with an alcohol and a catalyst to form a dialkyl terephthalate.
        - The terephthalate diester can be used with ethylene glycol to form a new polyester polymer.

    - Thus, it is possible to make the pure polymer again.

- Waste Plastic Pyrolysis to Fuel Oil

    - Another process involves conversion of assorted polymers into petroleum by a much less precise thermal depolymerization process.

    - Such a process would be able to accept almost any polymer or mix of polymers, including thermoset materials such as vulcanized rubber tires and the biopolymers in feathers and other agricultural waste.
    - Like natural petroleum, the chemicals produced can be used as fuels or as feedstock.

    - A RESEM Technology plant of this type in Carthage, Missouri, USA, uses turkey waste as input material.

- Gasification is a similar process, but is not technically recycling since polymers are not likely to become the result.

- Plastic Pyrolysis can convert petroleum-based waste stream such as plastics into quality fuels and carbons.

- Given below is the list of suitable plastic raw materials for pyrolysis:
    - Mixed plastic (HDPE, LDPE, PE, PP, Nylon, Teflon, PS, ABS, FRP, etc.)
    - Mixed waste plastic from waste-paper mill
    - Multi Layered Plastic

- Recycling Codes

    - In order to meet recyclers' needs while providing manufacturers a consistent, uniform system, a coding system is developed.

    - The recycling code for plastics was introduced in 1988 by plastics industry through the Society of the Plastics Industry, Inc.

    - Because municipal recycling programs traditionally have targeted packaging – primarily bottles and containers – the resin coding system offered a means of identifying the resin content of bottles and containers commonly found in the residential waste stream, depending on the type of resin.

        - <u>Type 1 plastic, PET (or PETE)</u>: polyethylene terephthalate, is commonly found in soft drink and water bottles.
        - <u>Type 2, HDPE</u>: high-density polyethylene is found in most hard plastics such as milk jugs, laundry detergent bottles, and some dishware.
        - <u>Type 3, PVC or V (vinyl)</u>: includes items like shampoo bottles, shower curtains, hoola hoops, credit cards, wire jacketing, medical equipment, siding, and piping.
        - <u>Type 4, LDPE</u>: low-density polyethylene, is found in shopping bags, squeezable bottles, tote bags, clothing, furniture, and carpet.
        - <u>Type 5, PP</u>: polypropylene, makes up syrup bottles, straws, Tupperware, and some automotive parts.
        - <u>Type 6, PS</u>: polystyrene, makes up meat trays, egg cartons, clamshell containers, and compact disc cases.
        - <u>Type 7</u> includes all other plastics like bulletproof materials, 3- and 5-gallon water bottles, and sunglasses.

    - Types 1 and 2 are the most commonly recycled.

Cost-Benefit Analysis

- There is some debate over whether recycling is economically efficient.

- It is said that dumping 10,000 tons of waste in a landfill creates six jobs, while recycling 10,000 tons of waste can create over 36 jobs.

- However, the cost effectiveness of creating the additional jobs remains unproven.

- According to the U.S. Recycling Economic Informational Study, there are over 50,000 recycling establishments that have created over a million jobs in the US.

    - Two years after New York City declared that implementing recycling programs would be "a drain to the city," New York City leaders realized that an efficient recycling system could save the city over $20 million.

- Municipalities often see fiscal benefits from implementing recycling programs, largely due to the reduced landfill costs.

    - A study conducted by the Technical University of Denmark according to the Economist, found that in 83% of cases, recycling is the most efficient method to dispose of household waste.

    - However, a 2004 assessment by the Danish Environmental Assessment Institute concluded that incineration was the most effective method for disposing of drink containers, even aluminum ones.

- Fiscal efficiency is separate from economic efficiency.

- Economic analysis of recycling does not include what economists call externalities, which are unpriced costs and benefits that accrue to individuals outside of private transactions.

    - Examples include:

        - ✓ Decreased air pollution and greenhouse gases from incineration
        - ✓ Reduced hazardous waste leaching from landfills
        - ✓ Reduced energy consumption
        - ✓ Reduced waste and resource consumption, which leads to a reduction in environmentally damaging mining and timber activity.

    - Without mechanisms such as taxes or subsidies to internalize externalities, businesses will ignore them despite the costs imposed on society.

    - To make such non-fiscal benefits economically relevant, advocates have pushed for legislative action to increase the demand for recycled materials.

    - The United States Environmental Protection Agency (EPA) has concluded in favor of recycling, saying that recycling efforts reduced the country's carbon emissions by a net 49 million metric ton in 2005.

- In the United Kingdom, the Waste and Resources Action Programme stated that Great Britain's recycling efforts reduce CO_2 emissions by 10-15 million tons a year.

- Recycling is more efficient in densely populated areas, as there are economies of scale involved.

- Certain requirements must be met for recycling to be economically feasible and environmentally effective.

- These include an adequate source of recyclates, a system to extract those recyclates from the waste stream, a nearby factory capable of reprocessing the recyclates, and a potential demand for the recycled products.

  - These last two requirements are often overlooked – without both an industrial market for production using the collected materials and a consumer market for the manufactured goods, recycling is incomplete and in fact only "collection."

- Many economists favor a moderate level of government intervention to provide recycling services.

- Economists of this mindset probably view product disposal as an externality of production and subsequently argue government is most capable of alleviating such a dilemma.

Trade-in Recyclates

- Certain countries trade in unprocessed recyclates.

- Some have complained that the ultimate fate of recyclates sold to another country is unknown and they may end up in landfills instead of reprocessed.

  - According to one report, in America, 50-80% of computers destined for recycling are actually not recycled.

  - There are reports of illegal waste imports to China being dismantled and recycled solely for monetary gain, without consideration for workers' health or environmental damage.

- - Although the Chinese government has banned these practices, it has not been able to eradicate them.

- Certain regions have difficulty using or exporting as much of a material as they recycle.

  - This problem is most prevalent with glass: both Britain and the U.S. import large quantities of wine bottled in green glass.

  - Though much of this glass is sent to be recycled, outside the American Midwest there is not enough wine production to use all of the reprocessed material.

  - The extra must be downcycled into building materials or reinserted into the regular waste stream.

- Similarly, the northwestern United States has difficulty finding markets for recycled newspaper, given the large number of pulp mills in the region as well as the proximity to Asian markets.

  - In other areas of the U.S., however, demand for used newsprint has seen wide fluctuation.

- In some U.S. states, a program called RecycleBank pays people to recycle, receiving money from local municipalities for the reduction in landfill space which must be purchased.

  - It uses a single stream process in which all material is automatically sorted.

Criticisms and Responses

- Complete recycling is impossible from a practical standpoint.

- In summary, substitution and recycling strategies only delay the depletion of non-renewable stocks and therefore may buy time in the transition to true or strong sustainability, which ultimately is only guaranteed in an economy based on renewable resources.

- Much of the difficulty inherent in recycling comes from the fact that most products are not designed with recycling in mind.

- The concept of sustainable design aims to solve this problem, and was laid out in the book Cradle to Cradle: Remaking the Way We Make Things by architect William McDonough and chemist Michael Braungart.

  - They suggest that every product (and all packaging they require) should have a complete "closed loop" cycle mapped out for each component – a way in which every component will either return to the natural ecosystem through biodegradation or be recycled indefinitely.

- While recycling diverts waste from entering directly into landfill sites, current recycling misses the dissipative components.

- Complete recycling is impracticable as highly dispersed wastes become so diluted that the energy needed for their recovery becomes increasingly excessive.

  - For example, how will it ever be possible to recycle the numerous chlorinated organic hydrocarbons that have bioaccumulated in animal and human tissues across the globe, the copper dispersed in fungicides, the lead in widely applied paints, or the zinc oxides present in the finely dispersed rubber powder that is abraded from automobile tires?

- As with environmental economics, care must be taken to ensure a complete view of the costs and benefits involved.

  - For example, paperboard packaging for food products is more easily recycled than most plastic, but is heavier to ship and may result in more waste from spoilage.

Energy and Material Flows

- The amount of energy saved through recycling depends upon the material being recycled and the type of energy accounting that is used.

- Correct accounting for this saved energy can be accomplished with life cycle analysis using real energy values.

  - In addition, exergy, which is a measure of useful energy, can be used.

- In general, it takes far less energy to produce a unit mass of recycled materials than it does to make the same mass of virgin materials.

- Some scholars use emergy (spelled with an m) analysis.

- Emergy calculations take into account economics which can alter pure physics-based results.

- Using emergy life-cycle analysis, researchers have concluded that materials with large refining costs have the greatest potential for high recycle benefits.

  - Moreover, the highest emergy efficiency accrues from systems geared toward material recycling, where materials are engineered to recycle back into their original form and purpose, followed by adaptive reuse systems where the materials are recycled into a different kind of product, and then by-product reuse systems where parts of the products are used to make an entirely different product.

- The Energy Information Administration (EIA) states on its website that "a paper mill uses 40% less energy to make paper from recycled paper than it does to make paper from fresh lumber."

- Some critics argue that it takes more energy to produce recycled products than it does to dispose of them in traditional landfill methods, since the curbside collection of recyclables often requires a second waste truck.

    - However, recycling proponents point out that a second timber or logging truck is eliminated when paper is collected for recycling, so the net energy consumption is the same.

- An emergy life-cycle analysis on recycling revealed that fly ash, aluminum, recycled concrete aggregate, recycled plastic, and steel yield higher efficiency ratios, whereas the recycling of lumber generates the lowest recycle benefit ratio.

- Hence the specific nature of the recycling process, the methods used to analyze the process, and the products involved affect the energy savings budgets.

- It is difficult to determine the amount of energy consumed or produced in waste disposal processes in broader ecological terms, where causal relations dissipate into complex networks or material and energy flow.

    - For example, cities do not follow all the strategies of ecosystem development.

    - Biogeochemical paths become fairly straight relative to wild ecosystems, with very reduced recycling, resulting in large flows of waste and low total energy efficiencies.

    - By contrast, in wild ecosystems, one population's wastes are another population's resources, and succession results in efficient exploitation of available resources.

    - However, even modernized cities may still be in the earliest stages of a succession that may take centuries or millennia to complete.

- How much energy is used in recycling also depends on the type of material being recycled and the process used to do so.

- Aluminum is generally agreed to use far less energy when recycled rather than being produced from scratch.

- The EPA states that recycling aluminum cans, for example, saves 95% of the energy required to make the same amount of aluminum from its virgin source, bauxite.

- In 2009 more than half of all aluminum cans produced came from recycled aluminum.

- Every year, millions of tons of materials are being exploited from the earth's crust and processed into consumer and capital goods.

    o After decades to centuries, most of the materials are "lost."
    o With the exception of some pieces of art or religious relics, they are no longer engaged in the consumption process. Where are they?

    o Recycling is only an intermediate solution for such materials, although it does prolong the residence time in the anthroposphere.

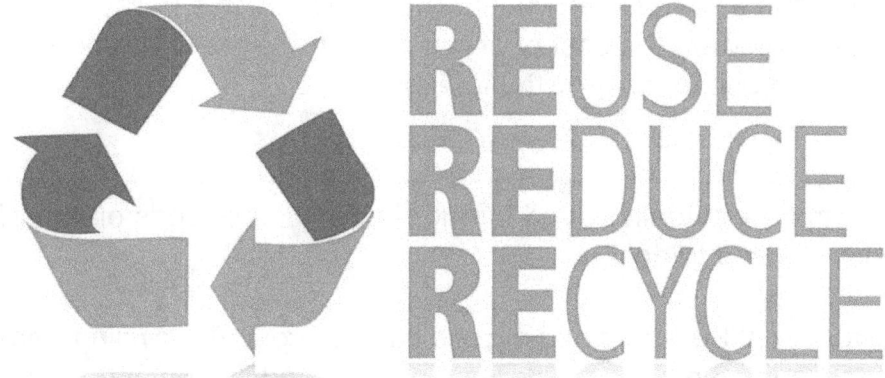

- For thermodynamic reasons, however, recycling cannot prevent the final need for an ultimate sink.

- Economist Steven Landsburg has suggested that the sole benefit of reducing landfill space is trumped by the energy needed and resulting pollution from the recycling process.

- Others, however, have calculated through life cycle assessment that producing recycled paper uses less energy and water than harvesting, pulping, processing, and transporting virgin trees.

    o When less recycled paper is used, additional energy is needed to create and maintain farmed forests until these forests are as self-sustainable as virgin forests.

- - Other studies have shown that recycling in itself is inefficient to perform the "decoupling" of economic development from the depletion of non-renewable raw materials that is necessary for sustainable development.

- As global consumption of a natural resource grows, its depletion is inevitable.

- The best recycling can do is to delay, complete closure of materials loops to achieve 100% recycling of non-renewables is impossible as micro-trace materials dissipate into the environment causing severe damage to the planet's ecosystems.

  - Historically, this was identified as the metabolic rift by Karl Marx, who identified the unequal exchange rate between energy and nutrients flowing from rural areas to feed urban cities that create effluent wastes degrading the planet's ecological capital, such as loss in soil nutrient production.

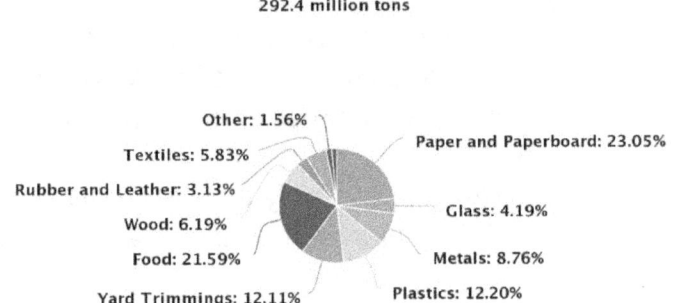

- Energy conservation also leads to what is known as Jevon's Paradox, where improvements in energy efficiency lowers the cost of production and leads to a rebound effect where rates of consumption and economic growth increases.

Costs

- The amount of money actually saved through recycling depends on the efficiency of the recycling program used to do it.

- The Institute for Local Self-Reliance argues that the cost of recycling depends on various factors around a community that recycles, such as landfill fees and the amount of disposal that the community recycles.

  - It states that communities start to save money when they treat recycling as a replacement for their traditional waste system rather than an add-on to it and by "redesigning their collection schedules and/or trucks."

- In some cases, the cost of recyclable materials also exceeds the cost of raw materials.

  - Virgin plastic resin costs 40% less than recycled resin.

- Additionally, a United States Environmental Protection Agency (EPA) study that tracked the price of clear glass from July 15 to August 2, 1991, found that the average cost per ton ranged from $40 to $60, while a USGS report shows that the cost per ton of raw silica sand from years 1993 to 1997 fell between $17.33 and $18.10.

- In a 1996 article for The New York Times, John Tierney argued that it costs more money to recycle the trash of New York City than it does to dispose of it in a landfill.

  - Tierney argued that the recycling process employs people to do the additional waste disposal, sorting, inspecting, and many fees are often charged because the processing costs used to make the end product are often more than the profit from its sale.

  - Tierney also referenced a study conducted by the Solid Waste Association of North America (SWANA) that found in the six communities involve in the study, "all but one of the curbside recycling programs, and all the composting operations and waste-to-energy incinerators, increased the cost of waste disposal."

  - Tierney also points out that "the prices paid for scrap materials are a measure of their environmental value as recyclables."

  - Scrap aluminum fetches a high price because recycling it consumes so much less energy than manufacturing new aluminum.

  - However, comparing the market cost of recyclable material with the cost of new raw materials ignores economic externalities – the costs that are currently not counted by the market.

  - Creating a new piece of plastic, for instance, may cause more pollution and be less sustainable than recycling a similar piece of plastic, but these factors will not be counted in market costs.

- A life cycle assessment can be used to determine the levels of externalities and decide whether the recycling may be worthwhile despite unfavorable market costs.

- Alternatively, legal means (such as a carbon tax) can be used to bring externalities into the market, so that the market cost of the material becomes close to the true cost.

- Critics also argue that while recycling may create jobs, they are often jobs with low wages and terrible working conditions.

  - These jobs are sometimes considered to be make-work jobs that do not produce as much as the cost of wages to pay for those jobs.

  - In areas without many environmental regulations and/or worker protections, jobs involved in recycling such as ship breaking can result in deplorable conditions for both workers and the surrounding communities.

Environmental Impact

- Economist Steven Landsburg, author of a paper entitled "Why I Am Not an Environmentalist," has claimed that paper recycling actually reduces tree populations.

    o He argues that because paper companies have incentives to replenish their forests, large demands for paper lead to large forests, while reduced demand for paper leads to fewer "farmed" forests.

- When foresting companies cut down trees, more are planted in their place.

    o Most paper comes from pulp forests grown specifically for paper production.

- Many environmentalists point out, however, that "farmed" forests are inferior to virgin forests in several ways.

- Farmed forests are not able to fix the soil as quickly as virgin forests, causing widespread soil erosion and often requiring large amounts of fertilizer to maintain while containing little tree and wildlife biodiversity compared to virgin forests.

- Also, the new trees planted are not as big as the trees that were cut down, and the argument that there will be "more trees" is not compelling to forestry advocates when they are counting saplings.

    o In particular, wood from tropical rainforests is rarely harvested for paper.
    o Rainforest deforestation is mainly caused by population pressure demands for land.

Possible Income Loss and Social Costs

- In some countries, recycling is performed by the entrepreneurial poor such as the *karung guni, zabbaleen*, the rag-and-bone man, waste picker, and junk man.

- With the creation of large recycling organizations that may be profitable, either by law or economies of scale, the poor are more likely to be driven out of the recycling and remanufacturing market.

- To compensate for this loss of income, a society may need to create additional forms of societal programs to help support the poor.

- In Brazil and Argentina, waste pickers/informal recyclers work alongside the authorities, in fully or semi-funded cooperatives allowing informal recycling to be legitimized as a paid public sector job.

- Because the social support of a country is likely to be less than the loss of income to the poor undertaking recycling, there is a greater chance the poor will come in conflict with the large recycling organizations.

- This means fewer people can decide if certain waste is more economically reusable in its current form rather than being reprocessed.

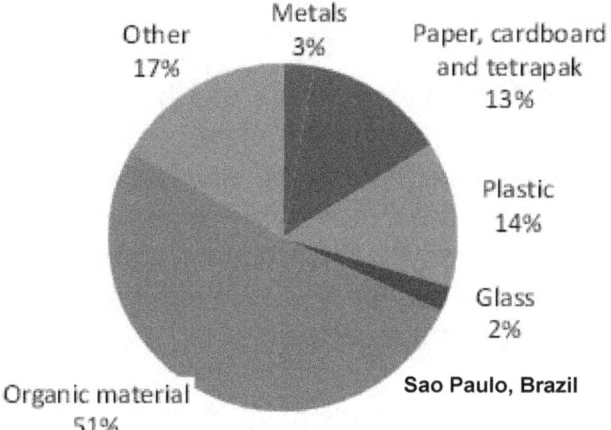

- Contrasted to the recycling poor, the efficiency of their recycling may actually be higher for some materials because individuals have greater control over what is considered "waste."

- One labor-intensive underused waste is electronic and computer waste.

  - Because this waste may still be functional and wanted mostly by those on lower incomes, who may sell or use it at a greater efficiency than large recyclers.

- Some recycling advocates believe that laissez-faire individual-based recycling does not cover all of society's recycling needs.

  - Thus, it does not negate the need for an organized recycling program.

- Local government can consider the activities of the recycling poor as contributing to property blight.

Public Participation in Recycling Programs

- Between 1960 and 2000, the world production of plastic resins increased 25-fold, while recovery of the material remained below 5%.

- Many studies have addressed recycling behavior and strategies to encourage community involvement in recycling programs.

- It has been argued that recycling behavior is not natural because it requires a focus and appreciation for long-term planning, whereas humans have evolved to be sensitive to short-term survival goals: and that to overcome this innate predisposition, the best solution would be to use social pressure to compel participation in recycling programs.

- However, recent studies have concluded that social pressure is unviable in this context.

  - One reason for this is that social pressure functions well in small group sizes of 50 to 150 individuals (common to nomadic hunter-gatherer peoples) but not in communities numbering in the millions, as we see today.

- o Another reason is that individual recycling does not take place in the public view.

- In a study it was found that personal contact with individuals within a neighborhood is the most effective way to increase recycling within a community.

  - o Other studies show that people who had friends and neighbors that recycled were much more likely to also recycle than those who didn't have friends and neighbors that recycled.

- Many schools have created recycling awareness clubs in order to give young students an insight on recycling.

- These schools believe that the clubs actually encourage students to not only recycle at school, but at home as well.

Additional Resources

Internet sites represent a vast resource of information. Using one of the search engines on the Internet such as Google or DuckDuckGo, find more information by searching for words, phrases, or websites: Always use caution when searching for information on the Internet. Follow guidelines to ensure the accuracy and reliability of the information you find.

Search words: recyclates, spectroscopic scanners, e-waste recycling

Self-Assessment

1. Identify recycling programs in your area. Are they widely used? Should the programs be expanded to include other items?

2. Of the seven types of plastic resin codes, which two are commonly recycled?

3. Of all the recycled products which do you feel impacts the environment the most and why?

4. How could you begin implementing more recycling in your home or school?

5. Describe the history of recycling.

30 Entomophagy

Major Concept

Entomophagy is the human use of insects as food – a practice common in human populations since prehistoric times. The practice is a relevant issue in the 21st century due to the rising cost of animal protein, food and feed security, environmental pressures, population growth, and increasing demand for protein among the middle classes.

Objectives

- Define entomophagy and provide examples of insects
- Describe the contribution of insects to food security
- List the environmental benefits of insect cultivation
- Identify the economic benefits insects as a food source

Key Terms

- Arachnids
- Arthropods
- Entomophagy
- Minilivestock
- Myriapods

Chapter Resource

Complementary *full color* illustrations, photos, charts, and graphs are available for Chapter 30 by following this URL: https://tinyurl.com/7vxzdaks This digital resource will enhance your understanding of the chapter concepts.

Entomophagy

- Entomophagy is the human use of insects as food.

- The eggs, larvae, pupae, and adults of certain insects have been eaten by humans from prehistoric times to the present day.

- Human insect-eating is common in cultures in most parts of the world, including North, Central and South America, Africa, Asia, Australia, and New Zealand.
  - Over 1,000 species of insects are known to be eaten in 80% of the world's nations.

  - The total number of ethnic groups recorded to practice entomophagy is around 3,000.

- However, in some societies, insect-eating is uncommon or even taboo.

- Today insect eating is rare in the developed world, but insects remain a popular food in many regions of Latin America, Africa, Asia, and Oceania.

- Some companies are trying to introduce insects into Western diets.

  - FAO has registered some 1900 edible insect species and estimates there were in 2005 some 2 billion insect consumers worldwide.

  - They also suggest entomophagy should be considered as a solution to environmental pollution.

Definition

- Entomophagy is sometimes defined broadly to cover the eating of arthropods other than insects, including arachnids and myriapods.

  - Insects and arachnids eaten around the world include:

    - ✓ Crickets
    - ✓ Cicadas
    - ✓ Grasshoppers
    - ✓ Ants
    - ✓ Various beetle grubs (such as mealworms)
    - ✓ The larvae of the darkling beetle or rhinoceros beetle
    - ✓ Various species of caterpillar (such as bamboo worms, mopani worms, silkworms, and waxworms)
    - ✓ Scorpions
    - ✓ Tarantulas

  - There are over 1,900 known species of arthropods that are edible to humans.

- Recent assessments of the potential of large-scale entomophagy have led some experts to suggest entomophagy as a potential alternative protein source to animal livestock, citing possible benefits including greater efficiency, lower resource use, increased food security, and environmental and economic sustainability.

Traditional Cultures

- A large number of cultures embrace the eating of insects.

- The species include 235 butterflies and moths, 344 beetles, 313 ants, bees, and wasps, 239 grasshoppers, crickets, and cockroaches, 39 termites, and 20 dragonflies, as well as cicadas.

- Insects are known to be eaten in 80% of the world's nations.

    - The leafcutter ant, *Atta laevigata*, is traditionally eaten in some regions of Colombia and northeast Brazil.
    - In southern Africa, the widespread moth *Gonimbrasia belina*'s large caterpillar, the mopani or mopane worm, is a source of food protein.
    - In Australia, the witchetty grub is eaten by the indigenous population.
    - The grubs of *Hypoderma tarandi*, a reindeer parasite, were part of the traditional diet of the Nunamiut people.
    - Use of insects as an ingredient in traditional foodstuffs in places such as Hidalgo in Mexico has been on a large enough scale to cause their populations to decline.

Western Culture

- Eating insects has not been adopted in the West, despite attempts to introduce insect-based food in Western markets.

- The most likely early adopters of insects as a meat substitute in Western societies have been profiled as younger males with a weak attachment to meat, who are open to trying novel foods and interested in the environmental impact of their food choice.

- Market introduction of insect as foods is usually done by small companies, often startups.

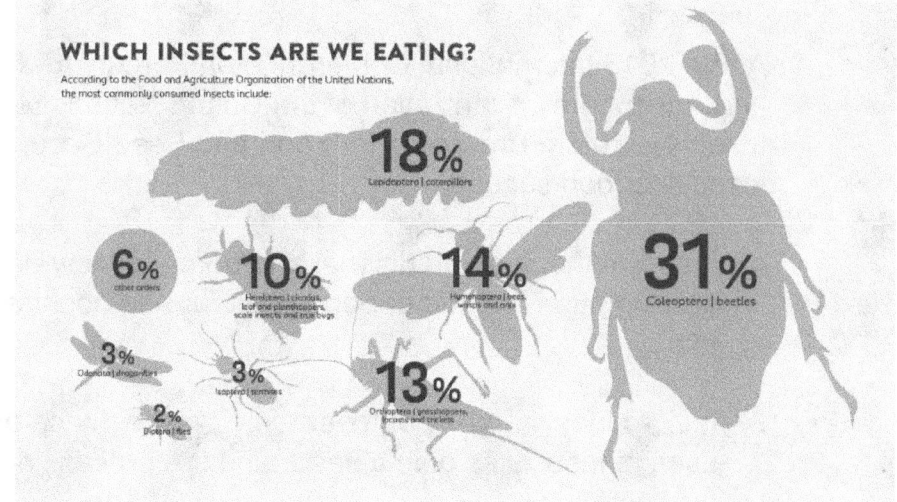

- A few companies have introduced products made using insects, whole or processed into food products.

    - Whole insects as snacks (Jimini's in France) or as novelties (HotLix lollipops in the US) are examples.

- In France, the first online shop for edible insects and products with edible insects came online in 2009.

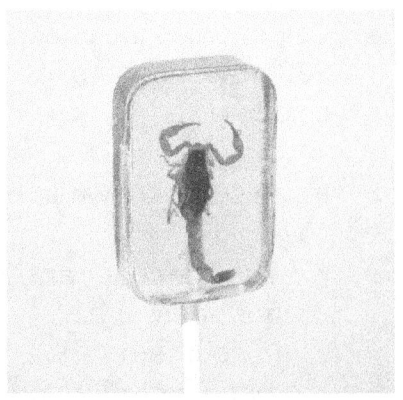

- They have been offered edible insects from Thailand, Africa, and Europe especially as an appetizer, a ready-to-cook range but also homemade candies and gluten-free protein bars with bugs.

- In 2011, Micronutris created the first French farm to raise insects.

- They grow mealworms and crickets in the South of France at Toulouse and craft fine products with them.

- In 2013, edible insects ready to cook are back with the first recipe book in France, "Délicieux ! 60 recettes à base d'insectes."

- The first North American edible insect online marketplace opened in 2015, and distributes products from entomophagy companies from the US, Mexico, Europe, and Thailand.

Food Security

- Insects as food and feed emerge as an especially relevant issue in the 21st century due to the rising cost of animal protein, food and feed security, environmental pressures, population growth, and increasing demand for protein among the middle classes.

- At the 2013 International Conference on Forests for Food Security and Nutrition, the Food and Agriculture Organization of the United Nations released a publication titled "Edible Insects – Future Prospects for Food and Feed Security" describing the contribution of insects to food security.

  - It shows the many traditional and potential new uses of insects for direct human consumption and the opportunities for and constraints to farming them for food and feed.

  - It examines the body of research on issues such as insect nutrition and food safety, the uses of insects as animal feed, and the processing and preservation of insects and their products.

- Minilivestock

    - The intentional cultivation of insects and edible arthropods for human food, referred to as minilivestock, is now emerging in animal husbandry as an ecologically sound concept.

    - Several analyses have found entomophagy to be a more environmentally friendly alternative to traditional animal livestock production.

    - In the Western world, agricultural technology companies such as Tiny Farms have been founded with the aim of modernizing insect rearing techniques, permitting the scale and efficiency gains required for insects to displace other animal proteins in the human food supply.

    - The first domestic insect farm, LIVIN Farms Hive, has recently been successfully started and will allow for the production of 200-500g of mealworm per week, a step toward a more distributed domestic production system.

- Therapeutic Foods

    - In 2012, Dr. Aaron T. Dossey announced that his company, All Things Bugs, had been named a Grand Challenges Explorations winner by the Bill & Melinda Gates Foundation.

    - Grand Challenges Explorations provides funding to individuals with ideas for new approaches to public health and development.

        - ✓ The research project is titled "Good Bugs: Sustainable Food for Malnutrition in Children."

    - Director of pediatric nutrition at the University of Alabama at Birmingham Frank Franklin has argued that since low calories and low protein are the main causes of death for approximately 5 million children annually, insect protein formulated into a ready-to-use therapeutic food similar to Nutriset's Plumpy'Nut could have potential as a relatively inexpensive solution to malnutrition.

    - In 2009, Dr. Vercruysse for Ghent University in Belgium proposed that insect protein can be used to generate hydrolysates, exerting both ACE inhibitory and antioxidant activity, which might be incorporated as multifunctional ingredients into functional foods.

        - ✓ Additionally, edible insects can provide a good source of unsaturated fats, thereby helping to reduce coronary disease.

- Indigenous Cultivation

    - Edible insects can provide economic, nutritional, and ecological advantages to the indigenous populations that commonly raise them.

        - ✓ For instance, the mopane worm of South Africa provides a "flagship taxon" for the conservation of mopane woodlands.

    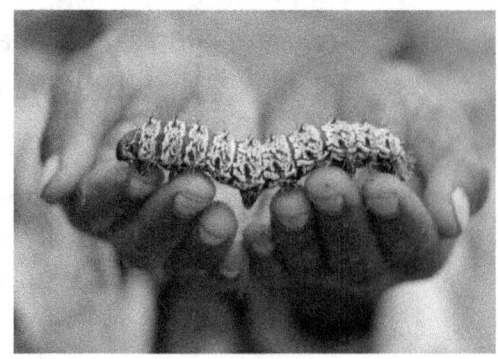

    - Some researchers have argued that edible insects provide a unique opportunity for insect conservation by combining issues of food security and forest conservation through a solution which includes appropriate habitat management and recognition of local traditional knowledge and enterprises.

    - However, senior FAO forestry officer Patrick Durst claims that *"Among forest managers, there is little knowledge or appreciation of the potential for managing and harvesting insects sustainably. On the other hand, traditional forest-dwellers and forest-dependent people often possess remarkable knowledge of the insects and their management."*

Environmental Benefits

- The methods of matter assimilation and nutrient transport used by insects make insect cultivation a more efficient method of converting plant material into biomass than rearing traditional livestock.

- More than 10 times more plant matter is needed to produce one kilogram of meat than one kilogram of insect biomass.

- The spatial usage and water requirements are only a fraction of that required to produce the same mass of food with cattle farming.

    - Production of 150g of grasshopper meat requires very little water, while cattle requires 3290 liters to produce the same amount of beef.

    - This indicates that lower natural resource use and ecosystem strain could be expected from insects at all levels of the supply chain.

- Edible insects also display exponentially faster growth and breeding cycles than traditional livestock.

- An analysis of the carbon intensity of five edible insect species conducted at the University of Wageningen, Netherlands found that "the average daily gain (ADG) of the five insect species studied was 4.0-19.6%, the minimum value of this range being close to the 3.2% reported for pigs, whereas the maximum value was 6 times higher. Compared to cattle (0.3%), insect ADG values were much higher."

- o Additionally, all insect species studied produced much lower amounts of ammonia than conventional livestock, though further research is needed to determine the long-term impact.

- o The authors conclude that insects could serve as a more environmentally friendly source of dietary protein.

Economic Benefits

- Insects generally have a higher food conversion efficiency than more traditional meats, measured as efficiency of conversion of ingested food, or ECI.

- While many insects can have an energy input to protein output ratio of around 4:1, raised livestock has a ratio closer to 54:1.

  - o This is partially due to feed first needs to be grown for most traditional livestock.

- Additionally, endothermic (warm-blooded) vertebrates need to use a significantly greater amount of energy just to stay warm whereas ectothermic (cold-blooded) plants or insects do not.

  - An index which can be used as a measure is the efficiency of conversion of ingested food to body substance.

    - ✓ For example, only 10% of ingested food is converted to body substance by feet cattle, versus 19-31% by silkworms and 44% by German cockroaches.

- Studies concerning the house cricket (*Acheta domesticus*) provide further evidence for the efficiency of insects as a food source.

  - o When reared at 30°C or more and fed a diet of equal quality to the diet used to rear conventional livestock, crickets showed a food conversion twice as efficient as pigs and broiler chicks, four times that of sheep, and six times higher than steers (oxen) when losses in carcass trim and dressing percentage are counted.

Additional Resources

Internet sites represent a vast resource of information. Using one of the search engines on the Internet such as Google or DuckDuckGo, find more information by searching for words, phrases, or websites: Always use caution when searching for information on the Internet. Follow guidelines to ensure the accuracy and reliability of the information you find.

Search words: minilivestock, therapeutic foods, indigenous cultivation, cricket flour

Top 7 Most Popular Mexican Insect Dishes
https://www.tasteatlas.com/most-popular-insect-dishes-in-mexico

Assessment

1. List five types of insects that are commonly eaten around the world.

2. Give a summary of your thoughts on using insects as a source of protein. Would you consume insects?

3. Name five insect products that are available in the U.S.

Glossary

The glossary for Just the Facts Essentials of Environmental Science can be found at this link https://tinyurl.com/7vxzdaks

www.ingramcontent.com/pod-product-compliance
Lightning Source LLC
Chambersburg PA
CBHW080243180526
45167CB00006B/2392